Tuberculous Meningitis

Tuberculous Meningitis

Manual of Diagnosis and Therapy

Edited by

JEROME HSI-CHENG CHIN

Department of Neurology, NYU Langone Health,
New York, NY, United States

ELSEVIER

ACADEMIC PRESS

An imprint of Elsevier

Academic Press is an imprint of Elsevier
125 London Wall, London EC2Y 5AS, United Kingdom
525 B Street, Suite 1650, San Diego, CA 92101, United States
50 Hampshire Street, 5th Floor, Cambridge, MA 02139, United States
The Boulevard, Langford Lane, Kidlington, Oxford OX5 1GB, United Kingdom

Notices

Knowledge and best practice in this field are constantly changing. As new research and experience broaden our understanding, changes in research methods, professional practices, or medical treatment may become necessary.

Practitioners and researchers must always rely on their own experience and knowledge in evaluating and using any information, methods, compounds, or experiments described herein. In using such information or methods they should be mindful of their own safety and the safety of others, including parties for whom they have a professional responsibility.

To the fullest extent of the law, neither the Publisher nor the authors, contributors, or editors, assume any liability for any injury and/or damage to persons or property as a matter of products liability, negligence or otherwise, or from any use or operation of any methods, products, instructions, or ideas contained in the material herein.

British Library Cataloguing-in-Publication Data

A catalogue record for this book is available from the British Library

Library of Congress Cataloging-in-Publication Data

A catalog record for this book is available from the Library of Congress

ISBN: 978-0-12-818825-5

For Information on all Academic Press publications
visit our website at https://www.elsevier.com/books-and-journals

Publisher: Nikki Levy
Acquisitions Editor: Melanie Tucker
Editorial Project Manager: Sara Pianavilla
Production Project Manager: Sujatha Thirugnana Sambandam
Cover Designer: Mark Rogers

Typeset by MPS Limited, Chennai, India

Contents

List of contributors

Alexandra Boubour
Department of Neurology, Columbia University Irving Medical Center, New York, NY, United States

Jerome H. Chin
Department of Neurology, NYU Langone Health, New York, NY, United States

Fiona V. Cresswell
Clinical Research Department, London School of Hygiene and Tropical Medicine, London, United Kingdom; Infectious Diseases Institute, College of Health Sciences, Makerere University, Kampala, Uganda; MRC-UVRI-LSHTM Uganda Research Unit, Entebbe, Uganda

Katarina M. Johansson Århem
Department of Infectious Diseases, Karolinska University Hospital, Stockholm, Sweden

Abdu K. Musubire
Infectious Diseases Institute, College of Health Sciences, Makerere University, Kampala, Uganda; Department of Medicine, School of Medicine, College of Health Sciences, Makerere University, Kampala, Uganda

Mandar Paradkar
BJ Government Medical College, Johns Hopkins University Clinical Research Site, Pune, India

Willy Ssengooba
Mycobacteriology Unit, Department of Medical Microbiology, Makerere University, Kampala, Uganda

Peter Ssenyonga
Department of Neurosurgery, Mulago National Referral Hospital, Kampala, Uganda

Kiran T. Thakur
Department of Neurology, Columbia University Irving Medical Center, New York, NY, United States

Preface

When I closed my private neurology practice in California, United States, to embark on a career in global/international health, I had no idea where I would be working and what diseases I would be treating. Eleven years later, I am the editor and a contributor to this first-ever medical book devoted to the diagnosis and treatment of tuberculous meningitis.

Tuberculosis is the oldest microbiologically confirmed infectious disease of humans and is now the leading infectious disease killer in the world. Tuberculosis is a global transmissible disease that can affect any person of any age. As a clinical neurologist teaching and treating patients in Africa, Asia, and the United States, I am keenly aware of the challenges of diagnosing and treating central nervous system infections, especially in health-care settings with limited laboratory services and constrained access to effective medications. The diagnosis and treatment of tuberculous meningitis is particularly challenging due to many factors, resulting in substantial mortality and morbidity.

I am deeply grateful to my colleagues and collaborators who have contributed their knowledge and experience to this endeavor. We have written this concise and clinically focused book to be a practical manual to assist and guide clinicians involved in evaluation and management of patients with neurological infections. My international work and this book could not have been possible without the love and support of my wife and two children. Lastly, I thank my patients and their families for the privilege of being their physician and for the trust they place in me to care for them to the best of my abilities.

Jerome Hsi-Cheng Chin

CHAPTER 1

Global and regional burden of tuberculosis and tuberculous meningitis

Alexandra Boubour[1], Mandar Paradkar[2] and Kiran T. Thakur[1]
[1]Department of Neurology, Columbia University Irving Medical Center, New York, NY, United States
[2]BJ Government Medical College, Johns Hopkins University Clinical Research Site, Pune, India

Key points

- Tuberculosis (TB) is a global disease and the leading infectious disease killer in the world.
- The highest incidence estimates for TB are in the World Health Organization African and Southeast Asia Regions.
- In low-burden countries the majority of TB cases are foreign-born persons from high-burden countries.
- TB is the leading cause of death for HIV-infected persons.
- Tuberculous meningitis is estimated to account for 1%—2% of all new cases of TB although reliable population-based data are limited.

Global tuberculosis epidemiology

Prevalence and incidence

Tuberculosis (TB) is the leading infectious cause of death on a global level, caused by *Mycobacterium tuberculosis* (MTB) [1,2]. According to World Health Organization (WHO) estimates, approximately 10 million (range, 9.0—11.1 million) incident cases of TB and 1.6 million TB deaths occurred in 2017, a small percentage decline from prior years (Fig. 1.1) [1]. Since 2000, global TB incidence has declined by 1.5% per year on average [1]. As of 2017, the incidence and number of TB cases remained greatest in the WHO Southeast Asia and African regions despite regional efforts for case reduction [1,3]. Nine percent of incident cases (920,000) occurred among HIV-positive people, 72% living in the African region [1]. Given these estimates, global TB incidence is not currently on track to meet the 2020 WHO End TB Strategy and United Nations (UN) Sustainable Development Goals (SDGs), which propose a 20% decrease in

Tuberculous Meningitis
DOI: https://doi.org/10.1016/B978-0-12-818825-5.00001-2

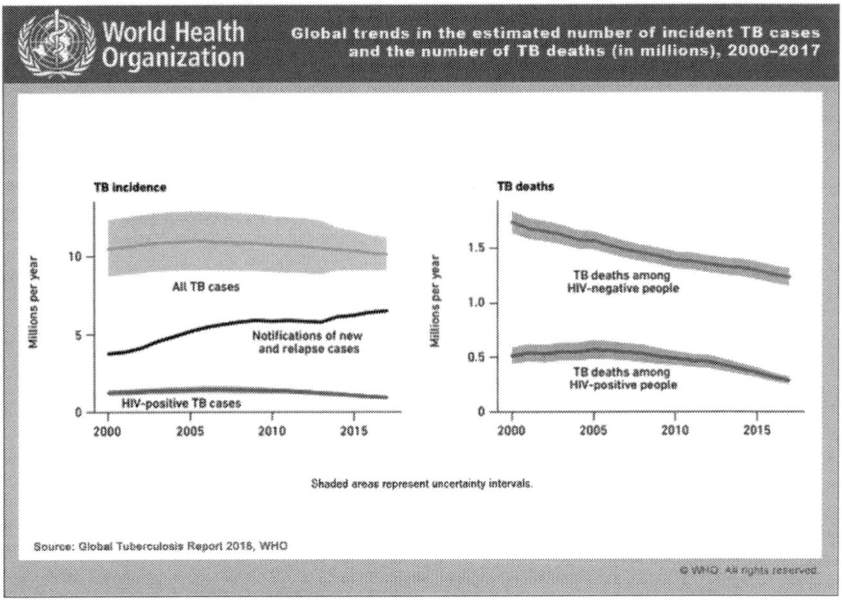

Figure 1.1 Global trends in the estimated number of incident TB cases and the number of TB deaths (in millions), 2000–2017.

incidence from 2015 [1,3,4]. This is complicated by the fact that just under two-thirds of new TB cases were reported in 2017, likely due to weak surveillance and registration systems in low-resource regions [1,5]. In 2017 the WHO estimated that 90% of cases were adults (≥15 years old), yet in endemic regions, children remain at highest risk [1].

Mortality

Globally, TB is the 10th leading cause of death overall and the leading cause of death in HIV-positive individuals, accounting for 40% of all deaths in this population (300,000 deaths; range, 266,000–335,000) [1]. TB case fatality rate (CFR) in 2017 was 15.7%; a drop from 23% in 2000. To align with the targets of the WHO End TB Strategy, the CFR must drop to 10% by 2020. Country CFRs range from <5% to >20%; most high-CFR countries are located in the WHO African Region, suggesting that many countries globally will not meet the WHO End TB Strategy goals [1,3]. In order to achieve TB eradication goals, TB prevention and treatment strategies must target the HIV-positive population to diminish the high incidence rates and mortality among this population. In the HIV-positive population, TB deaths decreased 44% from 2000 to 2017

(534,000–300,000) with an additional decline of 20% since 2015 [1]. Among HIV-negative persons, TB deaths have also declined from 1.8 million deaths in 2000 to 1.3 million deaths in 2017 (29% decline). TB deaths among HIV-negative people have decreased by an estimated 5% since 2015 (year 1 for the WHO End TB Strategy targets) [1]. The global TB mortality rate is decreasing 3% per year with an estimated 42% reduction from 2000 to 2017. The most rapid regional declines in mortality rates have occurred in the WHO European (11% yearly decline) and Southeast Asia regions (4% yearly decline) from 2013 to 2017 [1].

Regional tuberculosis epidemiology

A 2015 Global Burden of Disease (GBD) Tuberculosis Collaborators report stated that among HIV-negative individuals, mortality rates exceeded 50 per 100,000 population in Indonesia, Kiribati, Myanmar, Nepal, and 25 countries in sub-Saharan Africa [6]. Worldwide and in most regions, age-standardized TB prevalence, incidence, and mortality rates steadily drop with ascending sociodemographic index, a summary measure of a country or region's sociodemographic development [6]. However, the high incidence of TB in some Eastern European countries is the result of increasing HIV prevalence and insufficient systems for care and treatment [5]. Although TB and HIV are strongly related, TB incidence and mortality remain relatively high among HIV-negative people, particularly in Southeast Asia [6]. Conclusions of the 2015 GBD Study emphasize that countries with a high TB burden despite high sociodemographic development should look into reasons for high burden and formulate a rapid action plan [6]. Among countries with lower socioeconomic development, TB burden can be reduced through health systems strengthening to improve disease detection and provision of quality care, including access to essential TB medications [6].

High-burden tuberculosis settings

Globally, the WHO Southeast Asia and African regions make up an estimated 70% of all TB cases. Eight countries accounted for two-thirds of all TB cases in 2017, including India, Pakistan, Bangladesh, Indonesia, China, the Philippines, Nigeria, and South Africa (Fig. 1.2) [1]. Though the Southeast Asia region has higher total case numbers than the African region, likely due to relative population sizes, both regions had similar incidence rates in 2017 (Southeast Asia, 226 per 100,000; African,

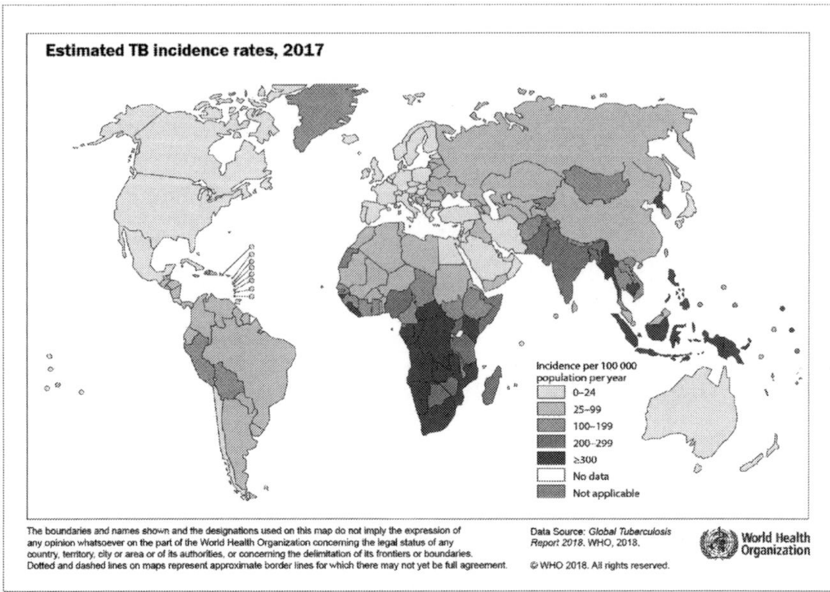

Figure 1.2 Estimated TB incidence rates, 2017.

237 per 100,000) [1]. Despite this, the percentage of TB cases with HIV coinfection in the African region (27%) was much greater than in the Southeast Asia region (3.5%) [1]. Following the WHO European Region, the WHO African Region had the steepest regional decline in incidence rates from 2013 to 2017 (5% in the European region and 4% in the African region per year on average) [1]. From 2013 to 2017, incidence rate reductions of 4%–8% per year occurred following an HIV epidemic peak and expanded TB–HIV prevention and care in southern Africa and following strengthened attempts to reduce TB burden in Russia (reductions of 5% per year) [1].

The outlook of national epidemics differs greatly. Per 100,000 population in 2017, there were <10 new cases in most high-income countries, 150–400 new cases in the 30 high-burden countries, and >500 new cases in Mozambique, the Philippines, and South Africa [1]. It is important to note that in countries where TB is endemic, the disease is fortified by weak surveillance and health systems, underreporting, and underdiagnosis [6]. Furthermore, in several TB-endemic countries, TB incidence has either plateaued or is declining more slowly than mortality, which is likely due to delays in treatment and diagnosis [6]. Despite global declines in incidence, in some countries, particularly in the WHO Southeast Asia and African Regions, population growth has resulted in an increasing absolute number of TB cases [5].

Low-burden tuberculosis settings

Although there is an unequal global distribution of TB, the disease remains a global epidemic with cases reported annually in every country. In 2018, 9029 new TB cases were reported in the United States (US) (0.7% decrease from 2017), and US TB incidence (2.8 per 100,000 people) decreased 1.3% from 2017 [7]. The incidence among non-US-born individuals was >14 times that in US-born individuals: among non-US-born individuals with TB, 2018 incidence was greatest among Asians and lowest among non-Hispanic Whites [7]. In England in 2017, 5102 new cases of TB were reported, the lowest number since 1990 [8]. People born outside of the United Kingdom accounted for 71% of TB cases, and the rate of TB among this population was 13 times higher than among those born in the United Kingdom [8]. In the US and the United Kingdom, populations at higher risk for TB include racial and ethnic minorities, homeless individuals, prisoners, and immunosuppressed individuals [5,7,8].

Drug-resistant tuberculosis

TB epidemiology is compounded by the rising prevalence of multidrug-resistant (MDR) (resistance to at least both rifampicin and isoniazid) and extensively drug-resistant (XDR) TB. In the future, this could increase global TB incidence, including in resource-rich and low-TB burden settings. As such, the WHO considers drug-resistant TB to be a global health security risk and public health crisis [1]. There were 558,000 new MDR TB or rifampin-resistant (RR) TB cases globally in 2017, which made up 5.6% of TB cases [9]. This amounts to 18% of previously treated cases and 3.5% of new TB cases with MDR TB [1]. MDR TB figures were highest in the Southeast Asia region: 192,000 (131,000–264,000) cases at a rate of 9.7 (6.7–13) per 100,000 population, followed by the Western Pacific and the European regions (Fig. 1.3) [1]. In 2017, 77 countries reported 10,800 cases of XDR TB; 88% of these countries were located in the WHO Southeast Asian and European regions [1]. Furthermore, the success rates for MDR and XDR TB treatment are as low as 55% and 34%, respectively, indicating very high mortality and morbidity associated with

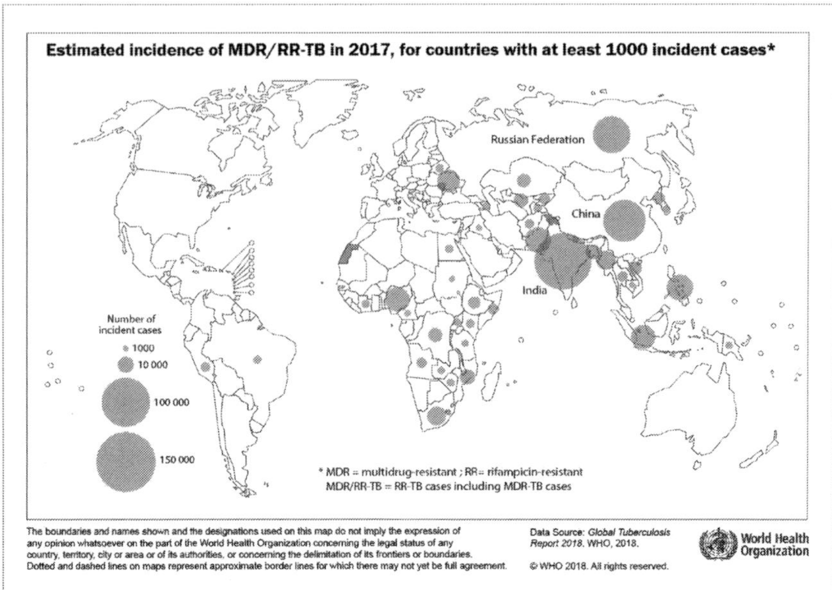

Figure 1.3 Estimated incidence of MDR/RR-TB in 2017, for countries with at least 1000 incident cases.

the disease- and treatment-induced toxicities [1]. In 2017 it is estimated that at least 230,000 people died of RR or MDR TB (CFR = 41%) [9].

Central nervous system tuberculosis

Central nervous system (CNS) involvement is the most severe TB manifestation, quoted to account for an estimated 5%–10% of extrapulmonary TB (EPTB) cases and 1% of all TB cases [5,10]. In 2017 there were an estimated 10 million new TB patients, of which 14% were diagnosed as EPTB [1]. Neurological manifestations of TB, including meningitis, spinal tubercular arachnoiditis, intracranial tuberculomas, and tuberculous brain abscesses, are five times more likely to occur in HIV-positive individuals than HIV-negative individuals [11,12]. Risk of mortality increases among patients who are older, HIV-positive, have hydrocephalus, and/or have a positive CNS-sourced MTB culture [13,14]. CNS TB patients have a higher risk of mortality than those without CNS involvement and disease severity is significantly associated with risk of death [12–16]. Although incidence and mortality of CNS TB have been consistently decreasing in the US and Canada, it is likely that regional and global CNS TB incidence and mortality remain underestimated due to underreporting and diagnostic difficulties [15,17,18].

In addition to HIV infection, childhood malnutrition, alcoholism, malignancies, and immunosuppressive agent use encompass the most significant risk factors for CNS TB [17]. A prospective study reported 824 cases of CNS TB (1% of all TB patients) in Canada from 1970 to 2001; CNS TB patients trended toward younger age, female sex, aboriginal identity, and foreign-born status compared with TB patients without CNS involvement [15]. Ghana- and Israel-based studies reported that a higher proportion of CNS TB patients were female [12,16]. A US-based study on 108 CNS TB cases found that 69.6% of cases were male, 42.4% of cases occurred among African-American patients, 40.3% of cases occurred among Hispanic patients, and 8.7% of cases occurred among Asian patients [13]. CNS TB cases in the US are most likely to occur among non-US-born individuals, particularly those of Hispanic ethnicity [7]. Studies in Oman and Israel have reported that CNS TB occurs more frequently in expatriates than in the local population [12,19].

The global incidence of tuberculous meningitis (TBM) is unknown, given that the absence of microbiological confirmation in many suspected TBM cases results in underreporting and difficulty in determining population-based estimates [10]. Among the reasons for underdiagnosis and undertreatment are the primary use of traditional healers, delayed access to medical care, underresourced medical and laboratory facilities, and low sensitivity cerebrospinal fluid culture and microscopy [5,20]. Regional studies have reported prevalence rates of TBM to be between 0.9% and 2.2% of all TB cases [8,21–23]. When TB prevalence is low in a population, most TBM cases occur in adults; however, in regions where TBM is endemic, the disease is more common among children, with children 6 months to 5 years old most likely to be affected [10]. Individuals in urban environments have a heightened risk of developing TB, including TBM [5,22]. TBM has become one of the most common etiologies of bacterial meningitis in endemic regions due to the efficacy and rollout of pneumococcal vaccines [24].

TBM is associated with high morbidity and mortality in all countries [6,13,15,17,18,25–29]. In 2016, 84 TBM cases were identified in the US, accounting for 4.2% of the reported 1882 EPTB cases and 0.9% of 9272 total TB cases [21]. Of those TBM patients, >50% died or had substantial neurological sequelae and other health complications. In the US, rates of neurological complications associated with TBM, including hydrocephalus, seizure, stroke, and audiovisual impairment, are increasing despite decreases in incidence and mortality [18]. A US-based study in Texas reported 192 patients with TBM (1.9% of all TB patients), of which 22.9% died, representing a mortality risk nearly five times greater than for non-TBM TB patients [21]. In 2017 death occurred among 15.7% ($N = 18$) of TBM cases in England [8]. A Brazil-based study reported mortality in 43.8% ($N = 56$) of TBM cases as well as an increasing incidence and a high prevalence of TBM among young adult White males [22]. A 2011 Cochrane review found median TBM mortality rates to be 41% without corticosteroids and 31% with corticosteroids [30]. Children have a greater risk of developing CNS TB than adults and in countries with low TB prevalence, children are at high risk of delayed diagnosis and poor outcome [31–33]. A systematic review and metaanalysis of treatment outcomes in 1636 pediatric TBM cases published a 19.3% mortality rate and reported that 53.9% of survivors experience neurological sequelae [26]. Children with TBM are significantly more likely than adults to endure long-term neurological sequelae [34,35].

Risk factors play a critical role in CNS TB incidence and control. For instance, alcohol misuse has been associated with poor EPTB prognosis, including treatment compliance and outcomes [6,28]. Additional risks for CNS TB include malignancies, childhood or adult measles, and long-time residence in a low- or middle-income country (LMIC) [36–41]. Children are more likely than adults to develop TBM from an in-home contact with pulmonary TB [35]. One study found that younger age and HIV infection were associated with CNS TB and CNS-sourced MTB culture positivity, and older age was associated with higher risk of death [13]. Additional studies have reported that older age, intravenous drug use, diabetes, chronic kidney failure, HIV infection, and positive MTB culture are independently associated with higher overall mortality among TBM patients [21,24,28,34].

Among adults, the most commonly reported risk factor for TBM is HIV coinfection. Due to immunosuppression, HIV-positive individuals are at increased risk of developing TB disease through acquisition or latent-to-active progression [42,43]. The proportion of HIV-associated meningitis cases caused by MTB is greater than one-half in TB-endemic regions [44]. A 2018 metaanalysis reported that 38% of all patients with TBM had HIV coinfection, and 6% of children with TBM had HIV coinfection [45]. In addition, a Kenyan-conducted autopsy study found that 26% of individuals with disseminated HIV–TB had undiagnosed

TBM [46]. A US-based study reported that approximately one-third of TBM patients were unaware of their positive HIV status and less than one-fifth of HIV-positive TBM patients were actively using antiretroviral therapy (ART) [29]. Patients with TBM–HIV coinfection are two to three times more likely to die than HIV patients without coinfection [29,44]. Within this population, overall mortality from TBM is approximately 40% (even for those prescribed ART), and most patients die during hospitalization [1,10,29].

World Health Organization management goals

Background

The WHO End TB Strategy and UN SDGs have created steep goals for reduction of TB in 2020–35, proposing a 35% decrease in TB mortality and 20% decrease in TB incidence by 2020 from 2015 [3,4]. However, to meet global targets for 2020–35, efforts must be augmented to improve TB prevention, diagnosis, and treatment. There are significant obstructions to TB eradication, including the rise of antimicrobial resistance, diagnostic and treatment challenges, the lack of robust health policy measures, and insufficient allocation of government funds for TB [1]. According to the WHO, despite slight global declines of estimated TB incidence and mortality in 2017, the decline in rates is insufficient to meet 2020 SDG targets [1,3]. The WHO and The Global Fund to Fight AIDS, Tuberculosis, and Malaria published that US$1.6 billion of international support was required each year to fill the 2014–16 TB control funding gap in 118 LMIC [3]. Since 2010, the TB developmental assistance growth rate has vastly declined, which complicates the urgent need to reduce TB burden in low-income countries [6].

Strides toward achieving management goals

Despite the major challenges facing the global population in the TB epidemic, strides in TB treatment provision have averted 49 million deaths worldwide from 2000 to 2015 [1]. One study found that a small national investment in social protection may result in a decreased burden of TB [47]. Increased gross domestic product allocation for social protection programs, which the International Labor Organization defines as "nationally defined sets of basic social security guarantees which secure protection aimed at preventing or alleviating poverty, vulnerability, and social exclusion," may lead to decreased TB morbidity and mortality [47,48].

For people at risk of developing TB and/or who may be TB-infected, TB preventive treatment (TPT) (typically daily isoniazid for $\geq$6 months) is most efficacious in TB prevention [9,49]. Currently, provision of TPT is recommended for HIV-positive people and to in-home contacts of patients with a clinical diagnosis of pulmonary TB [49,50]. Recent studies have demonstrated that short-course TPT treatment is just as effective as longer course treatment in HIV-positive individuals [51,52]. Sixty-seven countries published TPT use data among eligible HIV-positive persons and 138 countries reported on TPT use among children <5 years old in 2017 [9]. Among the reporting countries, an estimated 960,000 HIV-positive people received TPT (coverage estimation = 36%), which is not a substantial increase from the 930,000 reported in 2014 [9]. In recent years, TPT administration has nearly plateaued among HIV-positive people, which is likely due to reliance on ART [9,53]. Among people with HIV, ART is the most effective protective strategy against TB (including TBM) as it significantly reduces TBM risk regardless of purified protein derivative test status, antimycobacterial drug resistance, and baseline $CD4^+$ T-cell count [10,54]. ART reduces risk for TB acquisition and latent-to-active progression and decreases opportunistic infection incidence and mortality [55,56]. It must be noted that ART is only 65% effective at preventing TB among HIV-positive people; as mentioned, TB prevention efficacy can be improved with short-course TPT [9,51–53]. Current TPT coverage is well under the 2025 coverage target ($\geq$90%) set by the WHO End TB Strategy [3].

The Bacillus Calmette-Guérin (BCG) is a neonatal vaccine thought to be 73% effective in TBM prevention [1]. Recent studies have reported that 10–15-year vaccine effectiveness is approximately 51% in school-aged children and 40-year vaccine effectiveness is approximately 49% in adults [57,58]. BCG use is estimated to annually prevent 30,000 childhood cases of TBM [59]. Currently, some TB-endemic countries have a mandatory BCG vaccination policy and most recommend BCG vaccination [1]. Among countries with the highest TB burden, BCG vaccine varies widely from 53% to 99% coverage [1]. Childhood TBM can be prevented by administering isoniazid or other TPT to children who may have been exposed [9]. In 2017 nearly 292,000 eligible children <5 years old received TPT, which accounted for an estimated 23% of children eligible for TPT [9].

Challenges to achieving management goals

Despite overall declines in TB incidence, around 30% of the world population has latent TB [60]. These individuals are not clinically infected but carry a 10% lifetime risk for developing active TB [50]. Of the estimated 1.7 billion persons worldwide with latent TB infection (LTBI), it is estimated that 80% of cases are concentrated in the WHO Southeast Asia, Western Pacific, and African regions [60]. While the number of TB cases and incidence are the lowest ever reported in the US, studies using advanced statistical modeling have predicted that the US will not attain its TB elimination goal (incidence of <1 case per 1 million people annually) in the current century without dramatically increased investment in treatment and detection of LTBI [7]. Reactivation of remotely acquired LTBI is responsible for $>80\%$ of TB cases in the US; one US study found that 46.3% of 2018 TB cases in non-US-born individuals received a diagnosis more than 10 years following first US arrival [7].

Conclusion

There is a critical need to improve surveillance of CNS TB, the deadliest form of TB, which must be driven by scientific efforts to improve diagnosis and management of CNS TB [10]. It is postulated that CNS TB accounts for 1% of all TB cases, yet this is likely a significant underestimate [5,10]. Regional studies have reported prevalence rates of TBM to be between 0.9% and 2.2% of all TB cases, yet the external validity of these studies remains unknown [8,10,21–23]. Major public health efforts are required to define current incidence, prevalence, morbidity, and mortality of TBM, especially in light of the growing number of MDR TB cases worldwide. Increased health expenditures and structural interventions to alleviate social inequality and scale up preventive and treatment programs among high-risk populations are key elements to achieve goals set by the WHO and UN to reduce incidence and mortality rates for all TB [1,3,47,61].

References

[1] WHO. Available from: https://www.who.int/tb/publications/global_report/en/ Global tuberculosis report 2018. Geneva, Switzerland: World Health Organization; 2018 [accessed 03.10.19].

[2] Gagneux S. Ecology and evolution of *Mycobacterium tuberculosis*. Nat Rev Microbiol 2018;16(4):202–13.

[3] WHO. Available from: https://www.who.int/tb/post2015_strategy/en/The End TB Strategy. Geneva, Switzerland: World Health Organization; 2015 [accessed 03.10.19].
[4] UN. Available from: https://sustainabledevelopment.un.org/post2015/transformingourworld. Transforming our world: the 2030 Agenda for Sustainable Development. New York: United Nations; 2015 [accessed 03.10.19].
[5] Chin JH. Tuberculous meningitis: a neglected tropical disease? Neurol Clin Pract 2019;9(2):152–4.
[6] GBD Tuberculosis Collaborators. The global burden of tuberculosis: results from the Global Burden of Disease Study 2015. Lancet Infect Dis 2018;18(3):261–84.
[7] Talwar A, Tsang CA, Price SF, Pratt RH, Walker WL, Schmit KM, et al. Tuberculosis – United States, 2018. Am J Transplant 2019;19(5):1582–8.
[8] Public Health England. Available from: https://www.gov.uk/government/publications/tuberculosis-in-england-annual-report#historyTuberculosis in England: 2018 report. London, UK: Public Health England; 2018 [accessed 03.10.19].
[9] MacNeil A, Glaziou P, Sismanidis C, Maloney S, Floyd K. Global epidemiology of tuberculosis and progress toward achieving global targets – 2017. MMWR Morb Mortal Wkly Rep 2019;68(11):263–6.
[10] Wilkinson RJ, Rohlwink U, Misra UK, van Crevel R, Mai NTH, Dooley KE, et al. Tuberculous meningitis. Nat Rev Neurol 2017;13(10):581–98.
[11] Schaller MA, Wicke F, Foerch C, Weidauer S. Central nervous system tuberculosis: etiology, clinical manifestations and neuroradiological features. Clin Neuroradiol 2019;29(1):3–18.
[12] Daniele B. Characteristics of central nervous system tuberculosis in a low-incidence country: a series of 20 cases and a review of the literature. Jpn J Infect Dis 2014;67 (1):50–3.
[13] El Sahly HM, Teeter LD, Pan X, Musser JM, Graviss EA. Mortality associated with central nervous system tuberculosis. J Infect 2007;55(6):502–9.
[14] Jaipuriar RS, Garg RK, Rizvi I, Malhotra HS, Kumar N, Jain A, et al. Early mortality among immunocompetent patients of tuberculous meningitis: a prospective study. Am J Trop Med Hyg 2019;101(2):357–61.
[15] Phypers M, Harris T, Power C. CNS tuberculosis: a longitudinal analysis of epidemiological and clinical features. Int J Tuberc Lung Dis 2006;10(1):99–103.
[16] Ohene SA, Bakker MI, Ojo J, Toonstra A, Awudi D, Klatser P. Extra-pulmonary tuberculosis: a retrospective study of patients in Accra, Ghana. PLoS One 2019;14 (1):e0209650.
[17] Thakur K, Das M, Dooley KE, Gupta A. The global neurological burden of tuberculosis. Semin Neurol 2018;38(2):226–37.
[18] Merkler AE, Reynolds AS, Gialdini G, Morris NA, Murthy SB, Thakur K, et al. Neurological complications after tuberculous meningitis in a multi-state cohort in the United States. J Neurol Sci 2017;375:460–3.
[19] Gaifer Z. Epidemiology of extrapulmonary and disseminated tuberculosis in a tertiary care center in Oman. Int J Mycobacteriol 2017;6(2):162–6.
[20] Mitchell HK, Mokomane M, Leeme T, Tlhako N, Tsholo K, Ramodimoosi C, et al. Causes of pediatric meningitis in Botswana: results from a 16-year national meningitis audit. Pediatr Infect Dis J 2019;38(9):906–11.
[21] Nguyen DT, Agarwal S, Graviss EA. Trends of tuberculosis meningitis and associated mortality in Texas, 2010-2017, a large population-based analysis. PLoS One 2019;14 (2):e0212729.
[22] Souza CH, Yamane A, Pandini JC, Ceretta LB, Ferraz F, da Luz GD, et al. Incidence of tuberculous meningitis in the State of Santa Catarina, Brazil. Rev Soc Bras Med Trop 2014;47(4):483–9.

[23] Ducomble T, Tolksdorf K, Karagiannis I, Hauer B, Brodhun B, Haas W, et al. The burden of extrapulmonary and meningitis tuberculosis: an investigation of national surveillance data, Germany, 2002 to 2009. Euro Surveill 2013;18(12).
[24] Tsai KS, Chang HL, Chien ST, Chen KL, Chen KH, Mai MH, et al. Childhood tuberculosis: epidemiology, diagnosis, treatment, and vaccination. Pediatr Neonatol 2013;54(5):295−302.
[25] Banta JE, Ani C, Bvute KM, Lloren JIC, Darnell TA. Pulmonary vs. extra-pulmonary tuberculosis hospitalizations in the US [1998-2014]. J Infect Public Health. 2019. Available from: https://doi.org/10.1016/j.jiph.2019.07.001.
[26] Chiang SS, Khan FA, Milstein MB, Tolman AW, Benedetti A, Starke JR, et al. Treatment outcomes of childhood tuberculous meningitis: a systematic review and meta-analysis. Lancet Infect Dis 2014;14(10):947−57.
[27] Lee HG, William T, Menon J, Ralph AP, Ooi EE, Hou Y, et al. Tuberculous meningitis is a major cause of mortality and morbidity in adults with central nervous system infections in Kota Kinabalu, Sabah, Malaysia: an observational study. BMC Infect Dis, 16. 2016. p. 296.
[28] Qian X, Nguyen DT, Lyu J, Albers AE, Bi X, Graviss EA. Risk factors for extrapulmonary dissemination of tuberculosis and associated mortality during treatment for extrapulmonary tuberculosis. Emerg Microbes Infect 2018;7(1):102.
[29] Soria J, Metcalf T, Mori N, Newby RE, Montano SM, Huaroto L, et al. Mortality in hospitalized patients with tuberculous meningitis. BMC Infect Dis 2019;19(1):9.
[30] Marais BJ, Schaaf HS, Donald PR. Management of tuberculosis in children and new treatment options. Infect Disord Drug Targets 2011;11(2):144−56.
[31] Paulsrud C, Poulsen A, Vissing N, Andersen PH, Johansen IS, Nygaard U. Think central nervous system tuberculosis, also in low-risk children: a Danish nationwide survey. Infect Dis 2019;51(5):368−72.
[32] Cho YH, Ho TS, Wang SM, Shen CF, Chuang PK, Liu CC. Childhood tuberculosis in southern Taiwan, with emphasis on central nervous system complications. J Microbiol Immunol Infect 2014;47(6):503−11.
[33] Duque-Silva A, Robsky K, Flood J, Barry PM. Risk factors for central nervous system tuberculosis. Pediatrics 2015;136(5):e1276−84.
[34] Seddon JA, Shingadia D. Epidemiology and disease burden of tuberculosis in children: a global perspective. Infect Drug Resist 2014;7:153−65.
[35] Miftode EG, Dorneanu OS, Leca DA, Juganariu G, Teodor A, Hurmuzache M, et al. Tuberculous meningitis in children and adults: a 10-year retrospective comparative analysis. PLoS One 2015;10(7):e0133477.
[36] Klein NC, Damsker B, Hirschman SZ. Mycobacterial meningitis. Retrospective analysis from 1970 to 1983. Am J Med 1985;79(1):29−34.
[37] Ogawa SK, Smith MA, Brennessel DJ, Lowy FD. Tuberculous meningitis in an urban medical center. Medicine (Baltimore) 1987;66(4):317−26.
[38] Verdon R, Chevret S, Laissy JP, Wolff M. Tuberculous meningitis in adults: review of 48 cases. Clin Infect Dis 1996;22(6):982−8.
[39] Dube MP, Holtom PD, Larsen RA. Tuberculous meningitis in patients with and without human immunodeficiency virus infection. Am J Med 1992;93(5):520−4.
[40] Farer LS, Lowell AM, Meador MP. Extrapulmonary tuberculosis in the United States. Am J Epidemiol 1979;109(2):205−17.
[41] Yaramis A, Gurkan F, Elevli M, Soker M, Haspolat K, Kirbas G, et al. Central nervous system tuberculosis in children: a review of 214 cases. Pediatrics 1998;102(5):E49.
[42] Bucher HC, Griffith LE, Guyatt GH, Sudre P, Naef M, Sendi P, et al. Isoniazid prophylaxis for tuberculosis in HIV infection: a meta-analysis of randomized controlled trials. AIDS 1999;13(4):501−7.

[43] Selwyn PA, Hartel D, Lewis VA, Schoenbaum EE, Vermund SH, Klein RS, et al. A prospective study of the risk of tuberculosis among intravenous drug users with human immunodeficiency virus infection. N Engl J Med 1989;320(9):545–50.
[44] Gupta RK, Lucas SB, Fielding KL, Lawn SD. Prevalence of tuberculosis in post-mortem studies of HIV-infected adults and children in resource-limited settings: a systematic review and meta-analysis. AIDS 2015;29(15):1987–2002.
[45] Pormohammad A, Nasiri MJ, Riahi SM, Fallah F. Human immunodeficiency virus in patients with tuberculous meningitis: systematic review and meta-analysis. Trop Med Int Health 2018;23(6):589–95.
[46] Rana FS, Hawken MP, Mwachari C, Bhatt SM, Abdullah F, Ng'ang'a LW, et al. Autopsy study of HIV-1-positive and HIV-1-negative adult medical patients in Nairobi, Kenya. J Acquir Immune Defic Syndr 2000;24(1):23–9.
[47] Siroka A, Ponce NA, Lonnroth K. Association between spending on social protection and tuberculosis burden: a global analysis. Lancet Infect Dis 2016;16(4):473–9.
[48] International Labor Organization. The ILO social protection floors recommendation. Geneva, Switzerland: International Labor Organization; 2012. http://www.ilo.org/secsoc/areas-of-work/legal-advice/WCMS_205341/lang--en/index.htm; [accessed 03.10.19].
[49] Badje A, Moh R, Gabillard D, Guehi C, Kabran M, Ntakpe JB, et al. Effect of isoniazid preventive therapy on risk of death in west African, HIV-infected adults with high CD4 cell counts: long-term follow-up of the Temprano ANRS 12136 trial. Lancet Global Health 2017;5(11):e1080–9.
[50] WHO. Available from: https://www.who.int/tb/publications/2018/latent-tuberculosis-infection/en/Latent tuberculosis infection: updated and consolidated guidelines for programmatic management. Geneva, Switzerland: World Health Organization; 2018 [accessed 03.10.19].
[51] Swindells S, Ramchandani R, Gupta A, Benson CA, Leon-Cruz J, Mwelase N, et al. One month of rifapentine plus isoniazid to prevent HIV-related tuberculosis. N Engl J Med 2019;380(11):1001–11.
[52] Saunders MJ, Evans CA. Ending tuberculosis through prevention. N Engl J Med 2019;380(11):1073–4.
[53] Pathmanathan I, Ahmedov S, Pevzner E, Anyalechi G, Modi S, Kirking H, et al. TB preventive therapy for people living with HIV: key considerations for scale-up in resource-limited settings. Int J Tuberc Lung Dis 2018;22(6):596–605.
[54] Chaya S, Dangor Z, Solomon F, Nzenze SA, Izu A, Madhi SA. Incidence of tuberculosis meningitis in a high HIV prevalence setting: time-series analysis from 2006 to 2011. Int J Tuberc Lung Dis 2016;20(11):1457–62.
[55] Moreno S, Jarrin I, Iribarren JA, Perez-Elias MJ, Viciana P, Parra-Ruiz J, et al. Incidence and risk factors for tuberculosis in HIV-positive subjects by HAART status. Int J Tuberc Lung Dis 2008;12(12):1393–400.
[56] García de Olalla P, Martínez-González MA, Caylà JA, Jansà JM, Iglesias B, Guerrero R, et al. Influence of highly active anti-retroviral therapy (HAART) on the natural history of extrapulmonary tuberculosis in HIV patients. Int J Tuberc Lung Dis 2002;6(12):1051–7.
[57] Mangtani P, Nguipdop-Djomo P, Keogh RH, Sterne JAC, Abubakar I, Smith PG, et al. The duration of protection of school-aged BCG vaccination in England: a population-based case-control study. Int J Epidemiol 2018;47(1):193–201.
[58] Nguipdop-Djomo P, Heldal E, Rodrigues LC, Abubakar I, Mangtani P. Duration of BCG protection against tuberculosis and change in effectiveness with time since vaccination in Norway: a retrospective population-based cohort study. Lancet Infect Dis 2016;16(2):219–26.

[59] Trunz BB, Fine P, Dye C. Effect of BCG vaccination on childhood tuberculous meningitis and miliary tuberculosis worldwide: a meta-analysis and assessment of cost-effectiveness. Lancet 2006;367(9517):1173–80.
[60] Houben RM, Dodd PJ. The global burden of latent tuberculosis infection: a re-estimation using mathematical modelling. PLoS Med 2016;13(10):e1002152.
[61] Acevedo-Mendoza WF, Buitrago Gomez DP, Atehortua-Otero MA, Paez MA, Jimenez-Rincon M, Lagos-Grisales GJ, et al. Influence of socio-economic inequality measured by the Gini coefficient on meningitis incidence caused by *Mycobacterium tuberculosis* and *Haemophilus influenzae* in Colombia, 2008-2011. Infez Med 2017;25(1):8–12.

CHAPTER 2

Immunopathology of *Mycobacterium tuberculosis* complex

Willy Ssengooba[1] and Jerome H. Chin[2]
[1]Mycobacteriology Unit, Department of Medical Microbiology, Makerere University, Kampala, Uganda
[2]Department of Neurology, NYU Langone Health, New York, NY, United States

Key points

- *Mycobacterium tuberculosis* (MTB) genetic diversity influences the development of pulmonary and extrapulmonary tuberculosis.
- Tuberculous meningitis (TBM) accounts for a significant number of extrapulmonary tuberculosis cases and is associated with high rates of disability and mortality.
- MTB pathogen factors facilitate its survival in the host and dissemination to extrapulmonary sites.
- Host-specific factors are important determinants of susceptibility to and survival from TBM.

Globally, extrapulmonary tuberculosis (EPTB) accounted for 8%–24% of all notified cases of tuberculosis (TB) in 2017 [1]. Although vaccination with Bacillus Calmette–Guérin provides some protection against EPTB, these forms of TB still occur frequently, particularly among the pediatric population. EPTB carries a high risk of morbidity and mortality especially if it disseminates to the central nervous system, that is, tuberculous meningitis (TBM) and brain tuberculomas [2]. Early diagnosis of TBM and initiation of anti-TB treatment are essential to prevent serious neurological disability and death. It is believed that the pathogenesis of TBM begins with respiratory infection followed by early hematogenous dissemination to the meninges and the brain, indicating that understanding the host–pathogen interactive cascade is key to finding solutions for the prevention and improved treatment of TBM.

Biology of tuberculosis

From the era of Heinrich Herman Robert Koch, the founder of "modern" microbiology, including *Mycobacterium tuberculosis* (MTB) [3,4], the

Tuberculous Meningitis
DOI: https://doi.org/10.1016/B978-0-12-818825-5.00002-4

causative agent of TB, to the era of modern microbiology, the complex biology of MTB has challenged both scientists and clinicians. TB is a highly infectious disease that is transmitted from person to person through the air by tiny droplets (aerosols), 0.5–5.0 μm in diameter, expelled from an infected person. A person needs to inhale a dose of 1–200 bacilli to become infected. After infection the organism circumvents the natural body's immune system for its survival [5]. MTB bacilli travel to the alveoli of the lungs and are engulfed by the host alveolar macrophages where they multiply within the endosomes. Proinflammatory responses and recruitment of immune system cells lead to the formation of granulomas. Macrophages are intended to contain the infection, but if they fail the bacilli are released into the airway [5]. Although infection in the lungs occurs most commonly from inhalation of MTB (Ghon focus), pulmonary foci can also occur via hematogenous spread from another site (Simon focus) [5]. Hematogenous spread from the lungs to the brain and/or meninges leads to the formation of tuberculomas (Rich foci) that subsequently rupture and cause inflammatory meningitis [6]. Disseminated tuberculosis is more common in young children and HIV-infected individuals and is associated with diagnostic challenges and higher fatality rates [7–9]. Patients with impaired immunity are at risk for polyclonal infection as different strains of MTB may have distinct fitness in different host niches [10–13]. MTB infection may result in immediate clinical disease (pulmonary TB or EPTB) or become latent where MTB remains in the body without causing symptoms.

Mycobacterium tuberculosis complex and the pathogenesis of tuberculosis

MTB belongs to a seven-member complex [*Mycobacterium tuberculosis* complex (MTBC)] in which members have evolved from a common ancestor via successive DNA deletions resulting in differences in pathogenicity [14]. Global research teams have identified over a hundred potential virulence genes in MTBC [15]. A gene or protein is considered involved in virulence if its inactivation in the MTB genome leads to a measurable loss in pathogenicity or virulence in a validated TB model but fails to impair the bacterial growth in all standard in vitro conditions (except stress and starvation) in which the wild-type strain normally replicates. They include, but are not limited to, genes implicated in the adaptation of MTB to limited nutrient conditions and in mechanisms triggered by

MTB to counteract the microbicidal host cell responses. Several virulence factors have evolved in MTB in response to host immune interactions and are mostly geared toward MTB survival such as inhibiting the phagosome lysosome fusion to escape acidic environments [16]. MTB is unique compared to other bacterial pathogens due to a wide array of complex lipids and lipoglycans present in its cell wall. There is ample evidence that the cell wall is responsible for its survival, growth, permeability, virulence, and resistance to antibiotics, making it an ideal drug target [17]. Therefore to fully address the challenges of TB and EPTB, there is a need to study the factors facilitating the entry, survival, and dissemination of MTB in the host. This information is vital for the discovery of novel drug targets and for the development of more effective drugs and vaccines. MTB infection can persist in different body sites and cell types as reservoirs leading to delayed development of TB disease or even later reactivation to produce EPTB, with or without lung involvement. Unique pathogen factors may be responsible for the development of central nervous system TB.

Genetic diversity of *Mycobacterium tuberculosis* and tuberculous meningitis

A recent study showed the association of advanced immune suppression and increased prevalence of mixed-strain MTB infection [18] that may lead to selection of the most adapted strain to be disseminated. Whole-genome sequence analysis has revealed EPTB to be genotypically heterogeneous [12], and a study of *Mycobacterium avium* complex documented high genetic diversity in strains that cause pulmonary and disseminated disease [19]. Specific genotypes of MTB that cause extensive dissemination and brain infection in an experimental model have been previously documented [20]. While MTB strains in pulmonary TB and TB bacteremia have been found to be heterogeneous, especially among HIV-positive individuals [21,22], very little is known about the heterogeneity in TBM and other forms of EPTB. A study involving BALB/c mice infected with clinical isolates of MTB isolated from the cerebrospinal fluid (CSF) of a TBM patient showed rapid dissemination and brain infection [20,23]. The East Asian/Beijing lineage 2 is found to be associated with disease progression and CSF leukocyte count suggesting that the lineage may alter the presentation of meningitis by influencing the intracerebral inflammatory response [24]. The analysis of the genomes of MTB

Nonthaburi belonging to Indo-Oceanic lineage (lineage 1) isolated from TBM patients in Thailand revealed a novel genetic signature G2342203C [25]. The study further identified three mutations (T28910C, C1180580T, and C152178T) unique to the Nonthaburi isolates compared to genomes from the other seven lineages of MTBC [25]. This indicates that there are pathogen-specific mutations that could facilitate the penetration and replication of MTB in the meninges. These mutations may also be present in drug targets and lead to poor treatment outcomes if resistance is conferred against standard anti-TB drugs. A study done in Shaanxi province, China with 80% MTB Beijing genotype (lineage 2) revealed high rates of drug resistance among TBM cases [26]. Resistance to isoniazid, multidrug resistance, and MTB lineage have been found to be important determinants of mortality in patients with HIV-associated TBM [27]. HIV-infected patients with drug-resistant TBM have been documented to have severe clinical manifestations with exceptionally high mortality [28].

Human-adapted MTBC lineages demonstrate differences in clinical presentations, disease severity, and response to treatment [24]. Investigations of the MTB Beijing strains (lineage 2) in an animal model revealed high bacillary load and rates of dissemination [29]. Lineage 4, the Euro-American MTBC strains, was found to be less common in TBM compared to pulmonary TB [30,31]. In a study done in the United States, *Mycobacterium africanum* (lineage 6), endemic to West Africa, was found to be more likely to cause EPTB disease compared to MTB after controlling for HIV infection [32]. A recent study done in Mali found that *Mycobacterium africanum* shows slower sputum smear conversion on TB treatment than lineage 4 [33].

Host–pathogen interactions and the pathogenesis of tuberculous meningitis

Susceptibility to infection and disease are influenced by host-specific factors that modify TB pathogenesis. Studies of pathogens involved in meningeal infections have documented host factors, including the patient's phenotypes and genetic markers. Genome-wide association studies and human leukocyte antigen region fine-mapping studies have identified susceptibility loci for multiple common infections [34]. A study by Zhao et al. that analyzed genetic polymorphisms of CCL1 rs2072069 G/A and TLR2 rs3804099 T/C in pulmonary or meningeal TB patients found that

T597C polymorphism of TLR2 is a risk factor for susceptibility to PTB rather than to TBM [35]. Dunstan et al. found that rs17525495 is associated with susceptibility to bacteriologically confirmed bacterial meningitis but did not influence clinical presentation, disease severity, or survival following dexamethasone treatment [36]. Thuong et al. found that LTA4H genotype and HIV-infection influence pretreatment inflammatory phenotype and survival from TBM and further highlighted that LTA4H genotype may predict adjunctive corticosteroid responsiveness in HIV-uninfected individuals [37]. Studies have shown that strains of MTB Beijing genotype (lineage 2) and host genetics (polymorphisms in TLR2, TIRAP, and LTA4H genes) influence susceptibility to TBM [38]. Host-directed therapies tailored to patient LTA4H genotypes may prevent the detrimental effects of inflammatory responses to mycobacterial infection [39]. In a study done among a well-defined South African population, the mannose-binding protein B allele (G54D) was found to be protective against TBM [40].

There is a need to elucidate how host–pathogen interactions influence the pathogenesis of TBM. Ruesen et al. documented an association between homoplastic genetic variation in MTB genes and meningeal or pulmonary tuberculosis [41]. The study further revealed variations in the Rv0218 gene that encodes a secreted protein that could play a role in host–pathogen interactions by altering pathogen recognition or acting as a virulence effector. Polymorphisms in the macrophage receptor with collagenous structure that is vital in the phagocytosis of MTB have been associated with susceptibility and severity to pulmonary TB, a risk that is accelerated in infections involving MTB Beijing strains (lineage 2) [37]. Among TB patients in Vietnam the rs1052632 genotype GG of the major histocompatibility complex (MHC) Class I–like related molecule was strongly associated with the development of meningeal tuberculosis, and the absence of the G allele was associated with an increased risk of death with meningeal disease [42]. Genetic differences as a result of chromosomal rearrangements and protein globularity changes in MTB isolated from CSF compared to pulmonary samples highlight additional host–pathogen factors affecting TB dissemination [43].

Conclusion

TBM is the most lethal manifestation of TB and is associated with permanent neurological sequelae among many survivors. Our current understanding of the mechanisms facilitating the spread of MTB to the central nervous system and the development of inflammatory meningitis and tuberculomas is limited. Treatment outcomes using drug regimens developed for pulmonary TB are suboptimal (see Chapter 6: Treatment guidelines for tuberculosis and tuberculous meningitis). Therefore a more comprehensive understanding of host–pathogen interactions is needed to develop novel approaches for disease treatment, including host-directed therapies and patient-centered adjunctive therapies.

References

[1] WHO. Global tuberculosis report. World Health Organization, Geneva, Switzerland; 2018. <https://www.who.int/tb/publications/global_report/en/> [accessed 17.07.19].

[2] Chin JH. Neurotuberculosis: a clinical review. Semin Neurol 2019;39:1–6.

[3] Koch R. An address on the fight against tuberculosis in the light of the experience that has been gained in the successful combat of other infectious diseases. Br Med J 1901;2(2117):189–93.

[4] Koch R. A further communication on a remedy for tuberculosis. Br Med J 1891;1 (1568):125–7.

[5] Leung AN. Pulmonary tuberculosis: the essentials. Radiology 1999;210(2):307–22.

[6] Rock RB, Olin M, Baker CA, Molitor TW, Peterson PK. Central nervous system tuberculosis: pathogenesis and clinical aspects. Clin Microbiol Rev 2008;21:243–61.

[7] Jacob ST, Pavlinac PB, Nakiyingi L, Banura P, Baeten JM, Morgan K, et al. *Mycobacterium tuberculosis* bacteremia in a cohort of HIV-infected patients hospitalized with severe sepsis in Uganda-high frequency, low clinical suspicion [corrected] and derivation of a clinical prediction score. PLoS One 2013;8(8):e70305.

[8] Manabe YC, Nonyane BA, Nakiyingi L, Mbabazi O, Lubega G, Shah M, et al. Point-of-care lateral flow assays for tuberculosis and cryptococcal antigenuria predict death in HIV infected adults in Uganda. PLoS One 2014;9(7):e101459.

[9] Walters E, Duvenhage J, Draper HR, Hesseling AC, Van Wyk SS, Cotton MF, et al. Severe manifestations of extrapulmonary tuberculosis in HIV-infected children initiating antiretroviral therapy before 2 years of age. Arch Dis Child 2014;99 (11):998–1003.

[10] Adams LV, Kreiswirth BN, Arbeit RD, Soini H, Mtei L, Matee M, et al. Molecular epidemiology of HIV-associated tuberculosis in Dar es Salaam, Tanzania: strain predominance, clustering, and polyclonal disease. J Clin Microbiol 2012;50(8):2645–50.

[11] von Reyn CF, Kimambo S, Mtei L, Arbeit RD, Maro I, Bakari M, et al. Disseminated tuberculosis in human immunodeficiency virus infection: ineffective immunity, polyclonal disease and high mortality. Int J Tuberc Lung Dis 2011;15 (8):1087–92.

[12] Das S, Roychowdhury T, Kumar P, Kumar A, Kalra P, Singh J, et al. Genetic heterogeneity revealed by sequence analysis of *Mycobacterium tuberculosis* isolates from extra-pulmonary tuberculosis patients. BMC Genomics 2013;14:404.

[13] Perez-Lago L, Comas I, Navarro Y, Gonzalez-Candelas F, Herranz M, Bouza E, et al. Whole genome sequencing analysis of intrapatient microevolution in *Mycobacterium tuberculosis*: potential impact on the inference of tuberculosis transmission. J Infect Dis 2014;209(1):98–108.

[14] Ganeux S. Ecology and evolution of *Mycobacterium tuberculosis*. Nat Rev Microbiol 2018;16(4):202–13.
[15] Forrellad MA, Klepp LI, Gioffre A, Sabio y Garcia J, Morbidoni HR, et al. Virulence factors of the *Mycobacterium tuberculosis* complex. Virulence 2013;4 (1):3–66.
[16] Meena LS, Rajni. Survival mechanisms of pathogenic *Mycobacterium tuberculosis* H37Rv. FEBS J 2010;277(11):2416–27.
[17] Tahlan K, Wilson R, Kastrinsky DB, Arora K, Nair V, Fischer E, et al. SQ109 targets MmpL3, a membrane transporter of trehalose monomycolate involved in mycolic acid donation to the cell wall core of *Mycobacterium tuberculosis*. Antimicrob Agents Chemother 2012;56(4):1797–809.
[18] Shin SS, Modongo C, Ncube R, Sepako E, Klausner JD, Zetola NM. Advanced immune suppression is associated with increased prevalence of mixed-strain *Mycobacterium tuberculosis* infections among persons at high risk for drug-resistant tuberculosis in Botswana. J Infect Dis 2014;211(3):347–51.
[19] Uchiya K, Takahashi H, Yagi T, Moriyama M, Inagaki T, Ichikawa K, et al. Comparative genome analysis of *Mycobacterium avium* revealed genetic diversity in strains that cause pulmonary and disseminated disease. PLoS One 2013;8(8): e71831.
[20] Hernandez Pando R, Aguilar D, Cohen I, Guerrero M, Ribon W, Acosta P, et al. Specific bacterial genotypes of *Mycobacterium tuberculosis* cause extensive dissemination and brain infection in an experimental model. Tuberculosis 2010;90(4):268–77.
[21] Ssengooba W, de Jong BC, Joloba ML, Cobelens FG, Meehan CJ. Whole genome sequencing reveals mycobacterial microevolution among concurrent isolates from sputum and blood in HIV infected TB patients. BMC Infect Dis 2016;16:371.
[22] Ssengooba W, Cobelens FG, Nakiyingi L, Mboowa G, Armstrong DT, Manabe YC, et al. High genotypic discordance of concurrent *Mycobacterium tuberculosis* isolates from sputum and blood of HIV-infected individuals. PLoS One 2015;10(7): e0132581.
[23] Hernandez Pando R. Modelling of cerebral tuberculosis: hope for continuous research in solving the enigma of the bottom billion's disease. Malays J Med Sci 2011;18(1):12–15.
[24] Thwaites G, Caws M, Chau TT, D'Sa A, Lan NT, Huyen MN, et al. Relationship between *Mycobacterium tuberculosis* genotype and the clinical phenotype of pulmonary and meningeal tuberculosis. J Clin Microbiol 2008;46(4):1363–8.
[25] Coker OO, Chaiprasert A, Ngamphiw C, Tongsima S, Regmi SM, Clark TG, et al. Genetic signatures of *Mycobacterium tuberculosis* Nonthaburi genotype revealed by whole genome analysis of isolates from tuberculous meningitis patients in Thailand. PeerJ 2016;4:e1905.
[26] Wang T, Feng GD, Pang Y, Liu JY, Zhou Y, Yang YN, et al. High rate of drug resistance among tuberculous meningitis cases in Shaanxi province, China. Sci Rep 2016;6:25251.
[27] Tho DQ, Torok ME, Yen NT, Bang ND, Lan NT, Kiet VS, et al. Influence of antituberculosis drug resistance and *Mycobacterium tuberculosis* lineage on outcome in HIV-associated tuberculous meningitis. Antimicrob Agents Chemother 2012;56 (6):3074–9.
[28] Garg RK, Jain A, Malhotra HS, Agrawal A, Garg R. Drug-resistant tuberculous meningitis. Expert Rev Anti Infect Ther 2013;11(6):605–21.
[29] Tsenova L, Ellison E, Harbacheuski R, Moreira AL, Kurepina N, Reed MB, et al. Virulence of selected *Mycobacterium tuberculosis* clinical isolates in the rabbit model of meningitis is dependent on phenolic glycolipid produced by the bacilli. J Infect Dis 2005;192(1):98–106.

[30] Caws M, Thwaites G, Dunstan S, Hawn TR, Lan NT, Thuong NT, et al. The influence of host and bacterial genotype on the development of disseminated disease with *Mycobacterium tuberculosis*. PLoS Pathog 2008;4(3):e1000034.
[31] Click ES, Moonan PK, Winston CA, Cowan LS, Oeltmann JE. Relationship between *Mycobacterium tuberculosis* phylogenetic lineage and clinical site of tuberculosis. Clin Infect Dis 2012;54(2):211−19.
[32] Sharma A, Bloss E, Heilig CM, Click ES. Tuberculosis caused by *Mycobacterium africanum*, United States, 2004-2013. Emerg Infect Dis 2016;22(3):396−403.
[33] Diarra B, Kone M, Togo ACG, Sarro YDS, Cisse AB, Somboro A, et al. *Mycobacterium africanum* (Lineage 6) shows slower sputum smear conversion on tuberculosis treatment than *Mycobacterium tuberculosis* (Lineage 4) in Bamako, Mali. PLoS One 2018;13(12):e0208603.
[34] Tian C, Hromatka BS, Kiefer AK, Eriksson N, Noble SM, Tung JY, et al. Genome-wide association and HLA region fine-mapping studies identify susceptibility loci for multiple common infections. Nat Commun 2017;8(1):599.
[35] Zhao Y, Bu H, Hong K, Yin H, Zou YL, Geng SJ, et al. Genetic polymorphisms of CCL1 rs2072069 G/A and TLR2 rs3804099 T/C in pulmonary or meningeal tuberculosis patients. Int J Clin Exp Pathol 2015;8(10):12608−20.
[36] Dunstan SJ, Tram TT, Thwaites GE, Chau TT, Phu NH, Hien TT, et al. LTA4H genotype is associated with susceptibility to bacterial meningitis but is not a critical determinant of outcome. PLoS One 2015;10(3):e0118789.
[37] Thuong NT, Tram TT, Dinh TD, Thai PV, Heemskerk D, Bang ND, et al. MARCO variants are associated with phagocytosis, pulmonary tuberculosis susceptibility and Beijing lineage. Genes Immun 2016;17(7):41925.
[38] Brancusi F, Farrar J, Heemskerk D. Tuberculous meningitis in adults: a review of a decade of developments focusing on prognostic factors for outcome. Future Microbiol 2012;7(9):1101−16.
[39] Tobin DM, Roca FJ, Oh SF, McFarland R, Vickery TW, Ray JP, et al. Host genotype-specific therapies can optimize the inflammatory response to mycobacterial infections. Cell 2012;148(3):434−46.
[40] Hoal-Van Helden EG, Epstein J, Victor TC, Hon D, Lewis LA, Beyers N, et al. Mannose-binding protein B allele confers protection against tuberculous meningitis. Pediatr Res 1999;45(4 Pt 1):459−64.
[41] Ruesen C, Chaidir L, van Laarhoven A, Dian S, Ganiem AR, Nebenzahl-Guimaraes H, et al. Large-scale genomic analysis shows association between homoplastic genetic variation in *Mycobacterium tuberculosis* genes and meningeal or pulmonary tuberculosis. BMC Genomics 2018;19(1):122.
[42] Seshadri C, Thuong NT, Mai NT, Bang ND, Chau TT, Lewinsohn DM, et al. A polymorphism in human MR1 is associated with mRNA expression and susceptibility to tuberculosis. Genes Immun 2017;18(1):8−14.
[43] Saw SH, Tan JL, Chan XY, Chan KG, Ngeow YF. Chromosomal rearrangements and protein globularity changes in *Mycobacterium tuberculosis* isolates from cerebrospinal fluid. PeerJ 2016;4:e2484.

CHAPTER 3

Clinical presentations and features of tuberculous meningitis

Jerome H. Chin
Department of Neurology, NYU Langone Health, New York, NY, United States

Key points

- Tuberculosis begins as a pulmonary infection that can spread hematogenously to other parts of the body, including the brain and meninges.
- Tuberculous meningitis (TBM) is a subacute illness with symptoms developing for more than one week in most cases prior to medical evaluation.
- Meningitis symptoms such as neck stiffness and fever may be absent in TBM.
- Signs of dysfunction of the oculomotor and abducens nerves are common in TBM.
- Brain imaging reveals abnormalities in most cases of TBM with hydrocephalus being the most common finding.
- No diagnostic test or combination of tests can definitively rule out TBM.

This chapter is devoted to the clinical and radiological features of tuberculous meningitis (TBM). All of the photographs and images in the figures are of patients with definite TBM [positive nucleic acid amplification test and/or culture of *Mycobacterium tuberculosis* from cerebrospinal fluid (CSF)] except for the patient in Fig. 3.6. Detailed information on laboratory methods for detecting TBM and on medical and neurosurgical treatment of TBM can be found in Chapters 4–7 of this book.

Symptoms of tuberculous meningitis

Meningitis is an inflammation of the meninges that is comprised of three layers of tissue (dura mater, arachnoid mater, and pia mater) that cover and protect the brain and spinal cord. Within the meninges is the CSF that is produced continuously by the choroid plexus in the lateral, third, and fourth ventricles. Meningitis usually occurs in response to infections but

Tuberculous Meningitis
DOI: https://doi.org/10.1016/B978-0-12-818825-5.00003-6

can be triggered by noninfectious agents. Viral meningitis is considered a disease with low mortality and low risk of long-term neurological disability [1]. Bacterial meningitis, most commonly caused by *Streptococcus pneumoniae, Neisseria meningitidis, and Haemophilus influenzae*, is an acute severe illness that carries substantial risk of death and permanent neurological sequelae [2]. Fungal meningitis predominantly affects individuals with impaired immune defenses [3].

TBM is a form of extrapulmonary tuberculosis caused by hematogenous dissemination of *M. tuberculosis* from the lungs to the meninges (see Chapter 2: Immunopathology of *Mycobacterium tuberculosis* complex, for a discussion of pathogenic mechanisms). TBM is a subacute illness with symptoms progressing over 1–3 weeks before most patients seek medical attention. In a clinical trial of intensified TBM treatment conducted in Vietnam that enrolled 817 adults, the median duration of illness was 15 days [4]. Most patients with TBM are easily distinguished from patients with nonmycobacterial bacterial meningitis who present more acutely with severe symptoms. In a study of corticosteroids for bacterial meningitis conducted in Malawi that enrolled 465 adults, the median time to presentation in the placebo group was 72 hours (interquartile range, 48–144 hours) [5]. However, viral and fungal meningitis are often subacute and can be indistinguishable from TBM in the initial weeks of illness.

Early symptoms of TBM are headache, neck pain, vomiting, and fever. Notably, fever may be absent or low-grade. There may be recent symptoms to suggest pulmonary or disseminated tuberculosis; for example, weight loss, night sweats, cough, chest pain, and loss of appetite. Without treatment, patients with TBM invariably demonstrate more profound neurological symptoms after the first week of illness, which may include one or more of the following: impairment of consciousness, confusion, blurred vision, double vision, hemiparesis, paraparesis, and seizures. These symptoms are not expected in viral meningitis and, if present, should raise strong suspicion of TBM, fungal meningitis, or partially treated bacterial meningitis if the patient had already received antibacterial medications.

Clinical signs of tuberculosis and tuberculous meningitis

General physical examination of patients with tuberculosis and TBM may demonstrate fever, reduced breath sounds, adenopathy, hepatomegaly, and splenomegaly. Chest X-ray and/or chest CT studies should be obtained to look for evidence of pulmonary tuberculosis (Fig. 3.1). Neurological examination may reveal resistance and pain with passive flexion of the neck, that is, sign of meningeal irritation, depressed level of consciousness ranging from drowsiness to coma, dysconjugate gaze, papilledema, visual loss, hemiparesis, and paraparesis. Cranial nerve palsies, commonly of the oculomotor nerve (CN III) and/or the abducens nerve (CN VI) (Fig. 3.2), are noted in up to one-quarter of patients [6,7].

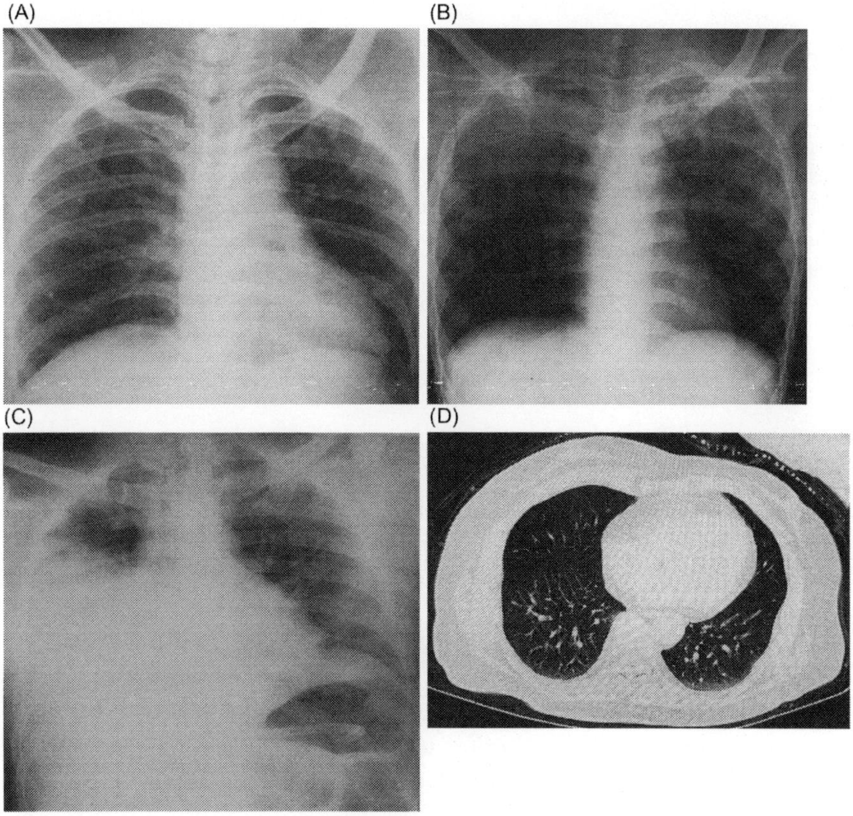

Figure 3.1 Chest X-rays (A–C) and chest computed tomographic image (D) of four patients with tuberculous meningitis. (A) Bilateral hilar adenopathy. (B) Bilateral upper lobe infiltrates. (C) Right lung consolidation with spontaneous pneumothorax. (D) Bilateral pulmonary nodules, bilateral pleural thickening, and pericardial effusion. *(D) By Dr. Dong-Hui Ao, Peking Union Medical College Hospital, Beijing, China.*

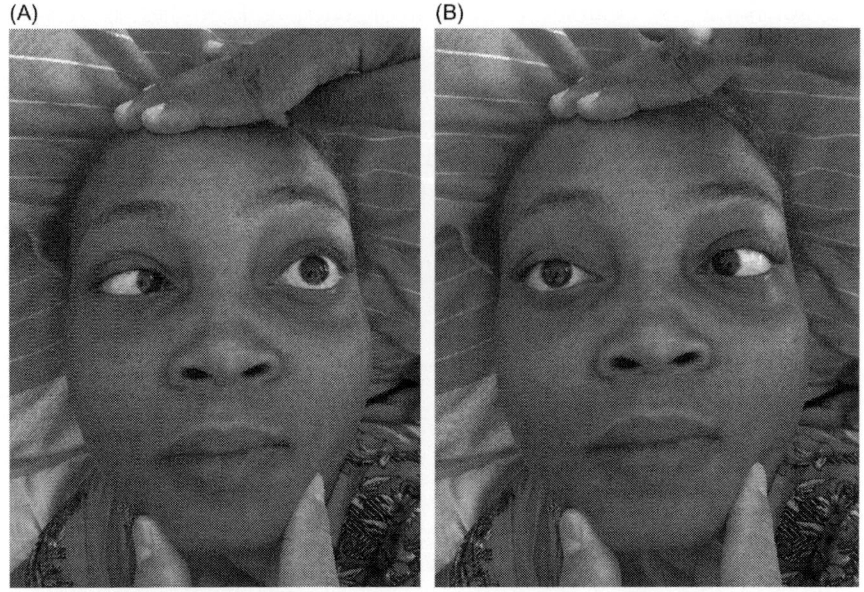

Figure 3.2 Bilateral palsies of cranial nerve VI (abducens nerve) in a patient with tuberculous meningitis presenting with a 1 week history of fever, headaches, neck pain, and vomiting. (A) Voluntary gaze to the left. (B) Voluntary gaze to the right.

Visual loss is difficult to identify in children and patients with depressed levels of consciousness. Visual acuity should be recorded in each eye as soon as the patient can cooperate with bedside testing using a handheld eye chart.

Cerebrospinal fluid of tuberculous meningitis

Lumbar puncture should be performed to obtain sufficient CSF for complete laboratory analyses, including white blood cell counts, protein and glucose levels, and microbiological studies (see Chapter 4: Laboratory methods for detecting tuberculosis and tuberculous meningitis). Opening CSF pressure is often elevated in the range of 25–40 cm of water. The CSF in TBM is typically clear or slightly turbid and straw-colored. Rarely, TBM can cause subarachnoid hemorrhage due to a necrotizing vasculitis of cerebral or spinal arteries leading to bloody and/or xanthochromic CSF. CSF white blood cell count is elevated with a lymphocytic predominance and protein levels are moderately high. CSF glucose level is usually but not always very low and can be measured at the bedside with a strip-based glucose meter (see Chapter 4: Laboratory methods for detecting tuberculosis and tuberculous meningitis).

Neuroimaging of tuberculous meningitis

Hydrocephalus is the enlargement of the ventricles of the brain due to obstruction of the normal flow of CSF through the ventricular system (noncommunicating) or to impairment of the resorption of CSF into the cerebral venous sinuses (communicating). Hydrocephalus is the most

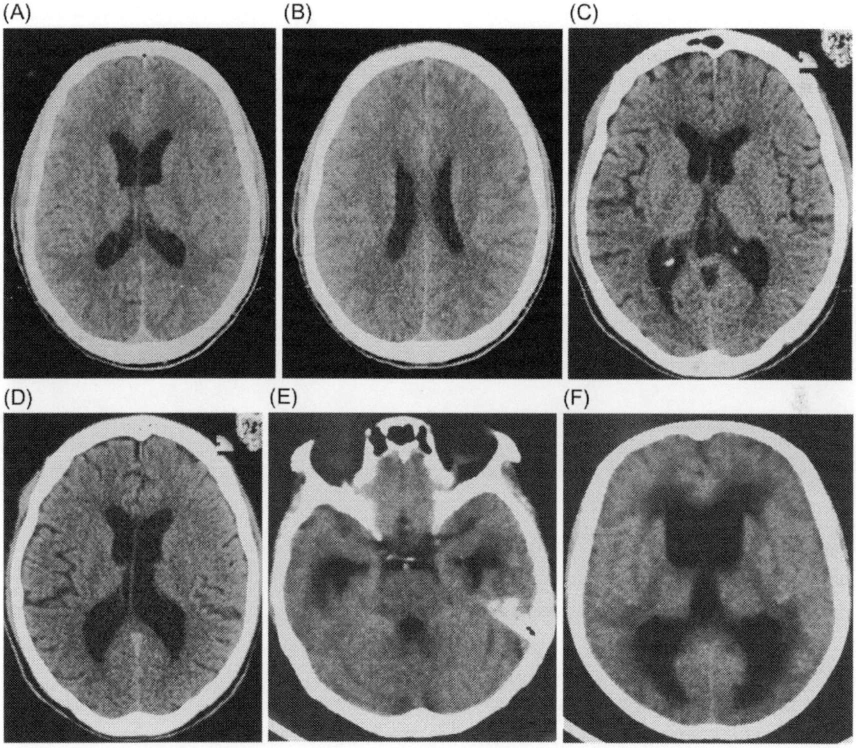

Figure 3.3 Hydrocephalus in three patients with tuberculous meningitis demonstrated on noncontrast computed tomographic images of the brain. (A and B) Mild hydrocephalus. (C and D) Moderate hydrocephalus. (E and F) Severe hydrocephalus. *(E and F) By Dr. Dong-Hui Ao, Peking Union Medical College Hospital, Beijing, China.*

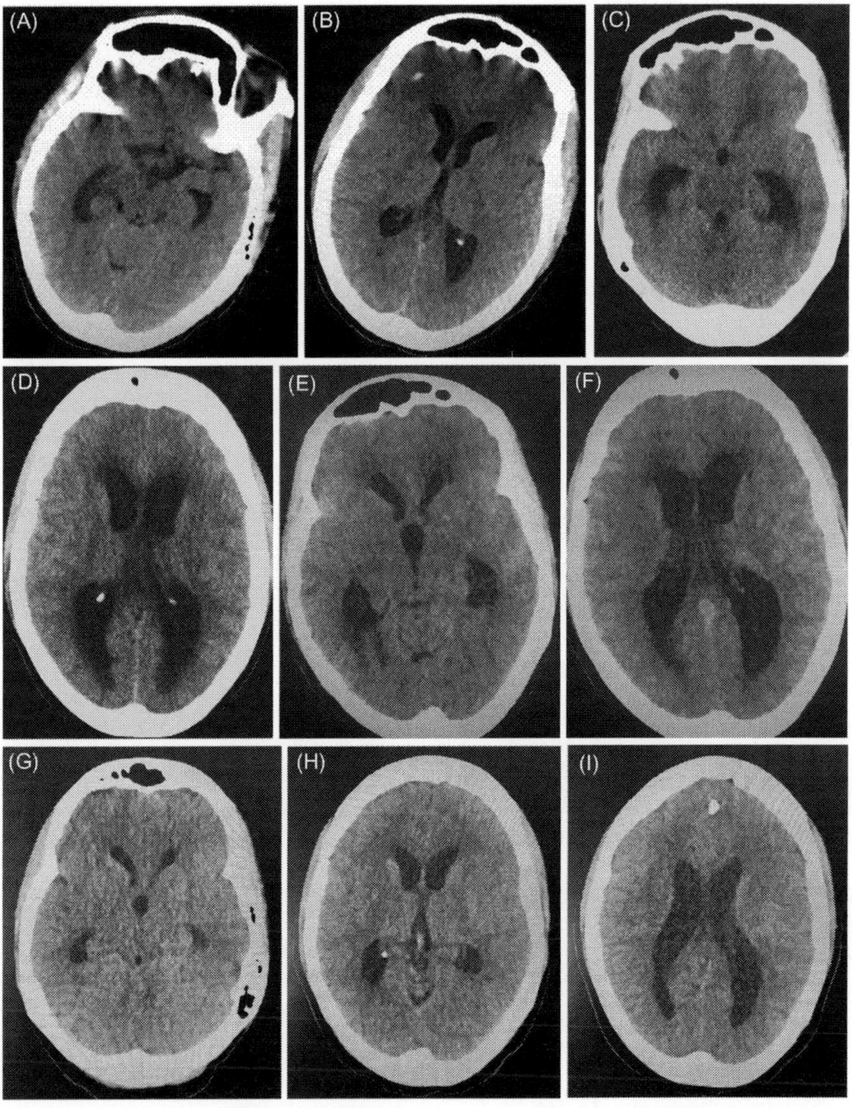

Figure 3.4 Noncontrast computed tomographic images of the brain of a patient with tuberculous meningitis showing the progression and improvement of hydrocephalus over 12 weeks. The patient was treated with acetazolamide and dexamethasone. (A and B) Week 1. (C and D) Week 3. (E and F) Week 7. (G–I) Week 12.

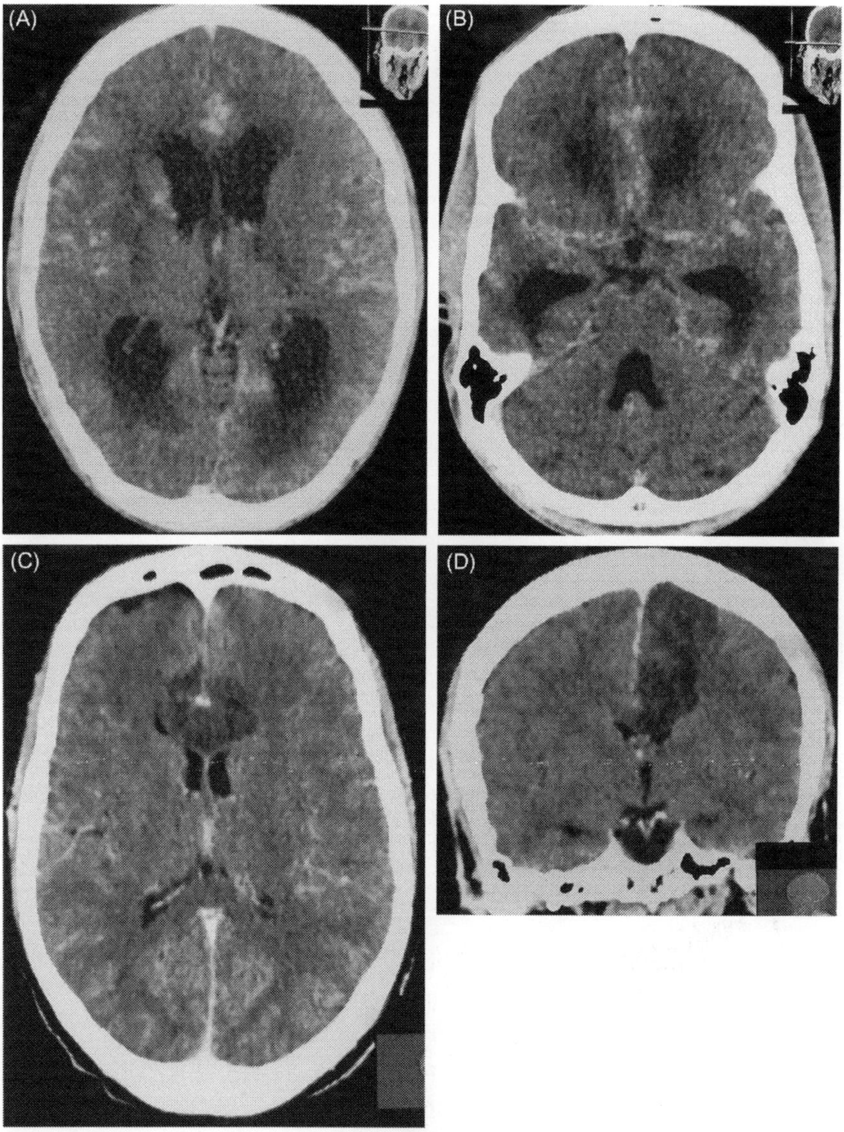

Figure 3.5 Cerebral infarcts in two patients with tuberculous meningitis. (A and B) Contrast-enhanced computed tomographic (CT) images of the brain showing diffuse and basal meningeal enhancement, hydrocephalus, and bilateral basal ganglia infarcts. (C and D) Contrast-enhanced CT images of the brain showing infarction in the vascular territory of the anterior cerebral artery.

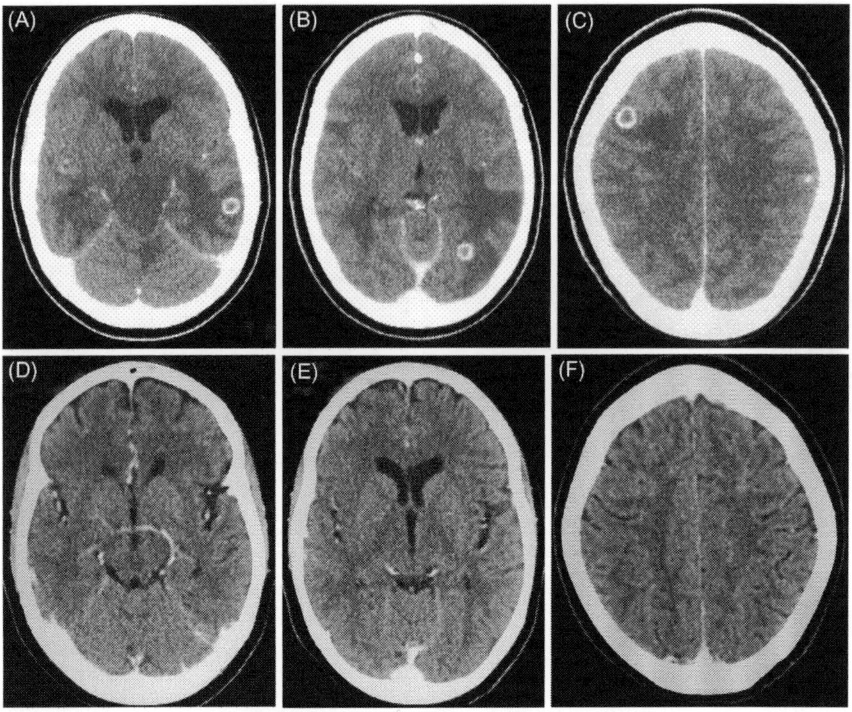

Figure 3.6 Multiple cerebral tuberculomas in a patient without meningitis. The patient presented with headaches, ataxia, and diplopia for 3 months (no seizures). Contrast-enhanced CT images of the brain show multiple ring-enhancing lesions with surrounding edema (A—C) that resolved completely after 6 months of standard antituberculosis therapy (D—F).

common abnormality seen on CT or MRI imaging of the brain in patients with TBM (Fig. 3.3) and can progress rapidly in the first few weeks of infection (Fig. 3.4). Treatment of hydrocephalus with medications (acetazolamide and/or furosemide) is often successful in neurologically stable patients (Fig. 3.4) but more severe cases will require neurosurgical intervention (see Chapter 7: Neurosurgical management of tuberculous meningitis).

Administration of intravenous contrast agents during CT or MRI imaging will often reveal robust enhancement of the meninges over the cerebral convexities as well as the base of the brain (Figs. 3.5 and 3.8). TBM may lead to a vasculitis involving the lenticulostriate arteries or major arteries of the circle of Willis leading to infarctions in the basal

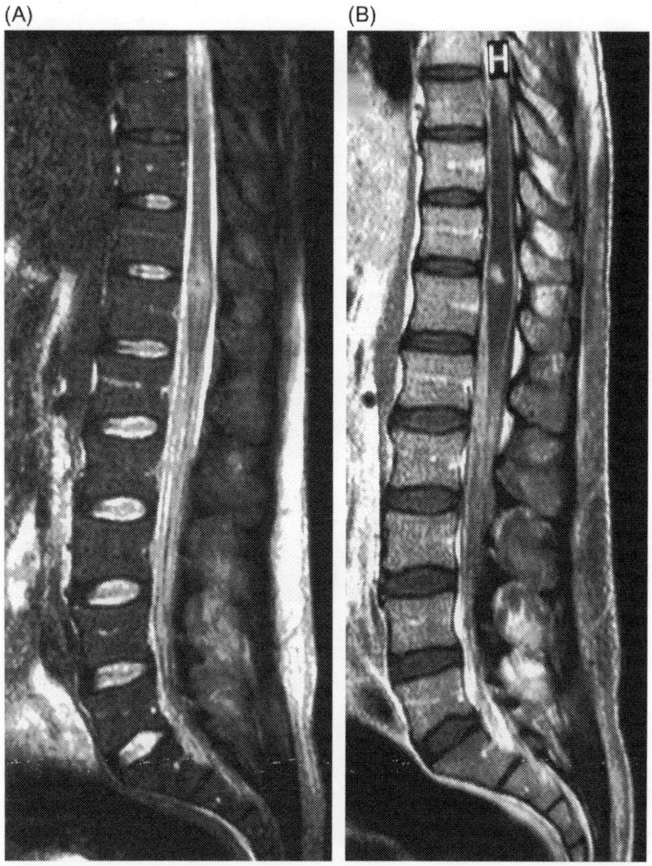

Figure 3.7 Myelitis and spinal cord tuberculoma in a patient with tuberculous meningitis. (A) T2-weighted sagittal magnetic resonance image of the lower spinal cord. (B) Contrast-enhanced T1-weighted sagittal image of the lower spinal cord.

ganglia or cerebral hemispheres (Fig. 3.5). Tuberculomas of the brain and spinal cord have an isointense or hypointense core with robust enhancement (homogeneous or ring-like) and moderate perilesional edema (Figs. 3.6–3.8). A distinctive feature of tuberculomas is the formation of conglomerated rings (Fig. 3.8). In patients with HIV coinfection, initiation of antiretroviral therapy may lead to clinical and radiological worsening with the development of new and/or enlarged cerebral tuberculomas (Fig. 3.8).

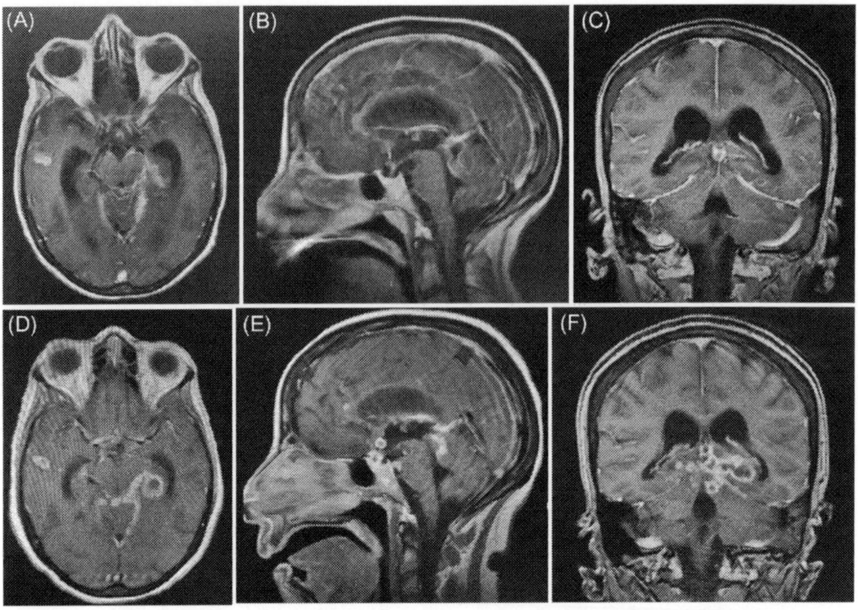

Figure 3.8 Immune reconstitution inflammatory syndrome in an HIV-infected patient with tuberculous meningitis (CD4 count = 125 cells/mm^3). (A–C) Contrast-enhanced magnetic resonance images of the brain showing diffuse and basal meningeal enhancement, hydrocephalus, and small tuberculomas. The patient was started on standard antituberculous therapy for drug-susceptible tuberculosis and initiation of antiretroviral therapy was delayed for 14 weeks. (D–F) Three weeks after initiation of antiretroviral therapy, multiple new tuberculomas developed in the cerebellum and temporal lobes. *By Osama Abu-Hadid, Rutgers New Jersey Medical School, Newark, New Jersey, United States.*

Conclusion

TBM is a subacute meningitis that should be strongly suspected in patients with a high risk of tuberculosis, for example, residents of and immigrants from high-burden countries, close contacts of known tuberculosis cases, and HIV-infected individuals. Meningitis patients with and without known risk factors for tuberculosis who demonstrate clinical and radiological features consistent with TBM and negative microbiological studies for nonmycobacterial bacteria and fungi should be treated presumptively for TBM. Microbiological confirmation of TBM in CSF is difficult to achieve; therefore negative rapid diagnostic tests and negative cultures for *M. tuberculosis* should not be used to rule out TBM or discontinue treatment with antituberculosis drugs.

References

[1] McGill F, Griffiths MJ, Bonnett LJ, Geretti AM, Michael BD, Beeching NJ, et al. Incidence, aetiology, and sequelae of viral meningitis in UK adults: a multicentre prospective observational cohort study. Lancet Infect Dis 2018;18:992–1003.

[2] McGill F, Heyderman RS, Panagiotou S, Tunkel AR, Solomon T. Acute bacterial meningitis in adults. Lancet 2016;388:3036–47.

[3] Schwartz S, Kontoyiannis DP, Harrison T, Ruhnke M. Advances in the diagnosis and treatment of fungal infections of the CNS. Lancet Neurol 2018;17:362–72.

[4] Heemskerk AD, Nguyen MTH, Dang HTM, Nguyen CVV, Nguyen LH, Do TDA, et al. Clinical outcomes of patients with drug-resistant tuberculous meningitis treated with an intensified antituberculosis regimen. Clin Infect Dis 2017;65:20–8.

[5] Scarborough M, Gordon SB, Whitty CJM, French N, Njalale Y, Chitani A, et al. Corticosteroids for bacterial meningitis in adults in sub-Saharan Africa. N Engl J Med 2007;357:2441–50.

[6] Hristea A, Olaru ID, Baicus C, Moroti R, Arama V, Ion M. Clinical prediction rule for differentiating tuberculous from viral meningitis. Int J Tuberc Lung Dis 2012;16:793–8.

[7] Heemskerk AD, Bang ND, Mai NTH, Chau TTH, Phu NH, Loc PP, et al. Intensified antituberculosis therapy in adults with tuberculous meningitis. N Engl J Med 2016;374:124–34.

CHAPTER 4

Laboratory methods for detecting tuberculosis and tuberculous meningitis

Jerome H. Chin[1] and Willy Ssengooba[2]
[1]Department of Neurology, NYU Langone Health, New York, NY, United States
[2]Mycobacteriology Unit, Department of Medical Microbiology, Makerere University, Kampala, Uganda

Key points

- Laboratory testing of cerebrospinal fluid (CSF) is useful to support a clinical suspicion of tuberculous meningitis (TBM).
- Increased lymphocyte counts, reduced glucose levels, and elevated protein levels are common CSF findings in TBM.
- Rapid diagnostic tests used for the identification of *Mycobacterium tuberculosis* (MTB) in CSF are smear microscopy and nucleic acid amplification tests (NAAT).
- Negative results on commercial or in-house NAAT cannot be used to exclude a diagnosis of TBM.
- MTB cultures require up to 6 weeks for the determination of growth. The decision to initiate treatment of suspected TBM should not be delayed while awaiting culture results.

The treatment of infectious diseases almost invariably begins with empiric antibiotic therapy to cover the most likely pathogens based on the clinical features and initial laboratory and radiologic studies. Microbiological or immunological confirmation of the causative bacteria, virus, fungus, or parasite may take days to weeks or in many cases, no pathogen will be identified. Further, in resource-limited settings without full laboratory facilities and diagnostic test kits, only the most basic testing will be available. Many nonmycobacterial bacteria that cause acute meningitis, for example, *Streptococcus pneumoniae*, *Neisseria meningitidis*, *Haemophilus influenzae*, can be seen on direct microscopy of cerebrospinal fluid (CSF) using Gram staining and can be detected using latex agglutination tests or rapid immunochromatographic diagnostic tests [1–3]. More sensitive but expensive rapid diagnostic platforms using polymerase chain reaction (PCR) technology are available to detect common viruses and nonmycobacterial bacteria in CSF [4,5].

Tuberculous Meningitis
DOI: https://doi.org/10.1016/B978-0-12-818825-5.00004-8

Mycobacterium tuberculosis (MTB) is a slow-growing bacterium with a doubling time of more than 17 hours in mice [6]. As described in Chapter 3, Clinical presentations and features of tuberculous meningitis (TBM), early diagnosis of TBM relies on a high clinical suspicion while ruling out other potential pathogens such as nonmycobacterial bacteria and fungi. TBM is often described as a paucibacillary disease due to the very low levels of MTB that can be detected in a sample of lumbar CSF. In this chapter, we review the currently available technologies for the rapid diagnosis of tuberculosis, both pulmonary and extrapulmonary, and the advantages and disadvantages of each methodology. Culture-based methods of MTB identification and associated drug susceptibility and drug resistance will be discussed in Chapter 5, Identification of *Mycobacterium tuberculosis* drug resistance.

Biosafety

Tuberculosis (TB) is an acquired disease through inhalation of infected aerosolized droplets containing MTB. Collection, handling, and processing liquid or semiliquid clinical specimens (sputum, blood, urine, pleural fluid, CSF) containing MTB carry a risk of exposure to the healthcare worker [7]. During lumbar puncture (LP) to collect CSF, the clinician performing the procedure should wear a mask covering the nose and mouth as well as glasses or goggles to protect the eyes. CSF under high pressure can exit the spinal needle in a spray or jet that can expose the clinician to infectious agents, including MTB and HIV. CSF should be collected carefully into tubes without touching the hub of the spinal needle. Contacting the hub of the spinal needle with the collection tube can cause the needle to flex and splash CSF upward when released.

According to the World Health Organization (WHO) biosafety manual for TB [7], the relative risk of laboratory-acquired TB for laboratory workers compared to nonlaboratory workers ranges from 1.4 for direct sputum smear microscopy to 7.8 for manipulating clinical specimens for culture. CSF collected from patients with possible TBM is usually processed for manual cell count, Gram stain, Ziehl–Neelsen stain or auramine stain, India Ink preparation, cryptococcal antigen, bacterial cultures, and additional tests as available to detect MTB, including liquid or solid media cultures. In addition, some laboratories routinely centrifuge CSF samples in an attempt to concentrate any MTB in the sample to increase diagnostic sensitivities. Centrifugation of CSF introduces significant risks of aerosolizing the samples. Sealed rotors, safety cups, and centrifuged specimen tubes should be opened in biological safety cabinets. Tubes that are vortexed or shaken to resuspend a pellet should be allowed to settle for at least 10 minutes before opening [7]. Manipulation of cultures for identification and drug-susceptibility testing carries a very high risk of infection.

Cerebrospinal fluid collection

Collection of CSF by LP is the most urgent and important procedure to perform in a case of suspected meningitis. In many middle- and high-income countries, commercial "LP kits" are available in the hospitals and contain all of the necessary equipment, including spinal needles, collection tubes, lidocaine, iodine-based disinfectant, glass or plastic tubes for measuring pressure, and sterile drapes. However, in most high-burden TB countries, particularly in sub-Saharan Africa and South/Southeast Asia, clinicians perform LPs using available intravenous cannula needles without local anesthesia and collect CSF in tubes used for blood sampling. If available, CSF should be collected in sterile plain tubes with secure screw caps [8]. For improvised measurement of CSF pressure, a sterile piece of intravenous fluid tubing may be used and the height of the CSF meniscus marked and measured with a ruler (Fig. 4.1). Normal CSF pressure is less than 200 mm of water in the lateral decubitus position with the patient relaxed. Three-to-four tubes should be filled with a minimum of 1.0 mL of CSF per tube. More should be collected if the CSF is flowing well (up to 5.0 mL per tube) to allow for additional diagnostic studies.

It should be noted that patients and families may refuse LP for fear of serious or fatal complications. In these cases, TBM will need to be diagnosed or excluded based on available clinical information (history, exam, comorbidities) and other diagnostic studies (see Chapter 3: Clinical presentations and features of tuberculous meningitis). Further, LP may be considered dangerous and not performed if there are clinical signs to suggest a risk for brain herniation, and neuroimaging (CT or MRI) is not available. Severe depression of level of consciousness (stupor, coma) in a case of suspected meningitis may indicate diffuse cerebral or brainstem ischemia, obstructive or communicating hydrocephalus, multiple infarcts, or brain abscesses with edema. TBM often causes third

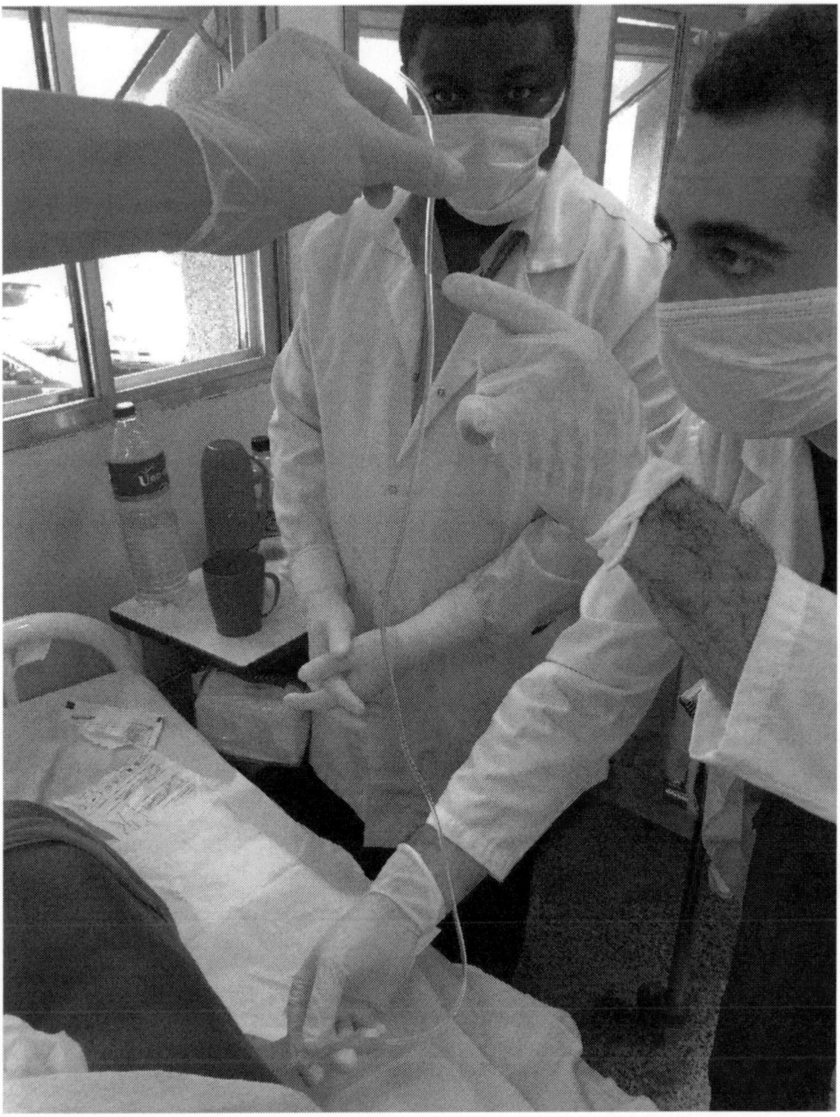

Figure 4.1 Checking CSF opening pressure with a piece of sterile intravenous line tubing. *CSF*, Cerebrospinal fluid.

and sixth cranial nerve palsies due to basal meningitis and vasculitis of the cranial nerves that could be interpreted as possible brain herniation in the absence of confirmatory neuroimaging. In this author's experience a conscious patient with a third or sixth cranial nerve palsy who is speaking and following commands without a hemiparesis is unlikely to have a herniation syndrome and can safely undergo an LP even if neuroimaging is not available. As a general rule, not performing an LP and obtaining CSF for testing will likely result in a higher mortality and morbidity risk for a patient with suspected meningitis than performing an LP.

Routine cerebrospinal fluid studies

Distinguishing different etiologies of meningitis often can be inferred from the results of basic CSF analyses, including glucose level, protein level, and microscopic manual white blood cell count and differential (Table 4.1) [2]. The CSF glucose level is perhaps the most informative test for suspected TBM and can be performed quickly and reliably during LP using a handheld inexpensive glucometer (Fig. 4.2). The test strip is contacted directly to a drop of CSF at the hub of the spinal needle for an immediate reading. Studies have shown very comparable results between CSF glucose measurements tested with a portable glucometer and tested in the laboratory biochemistry analyzer [9]. Blood glucose should be obtained after the LP by finger stick. Normal CSF glucose in adults is 50%–80% of blood glucose. CSF should be delivered at room temperature to the laboratory, and manual cell counts should be performed within 90 minutes of sampling since white blood cells will disintegrate beyond this time (normal CSF cell count $\leq$ 5 white blood cells/cubic mm) [10]. White and red blood cell counts should be performed on the third or fourth tube if multiple tubes are collected.

In TBM the CSF glucose is often but not always low (normal > 2.2 mmol/L). The CSF protein is usually elevated (normal 0.15–0.45 g/L), and the CSF cell count is moderately high with a lymphocytic predominance. The CSF opening pressure is typically elevated between 200 and 400 mm of water. In highly immunosuppressed patients with HIV coinfection, the CSF cell count may be normal [14]. It is important to note that the CSF profiles in viral, fungal, bacterial, and MTB meningitis overlap and cannot be used exclusively to diagnose the etiologic pathogen. All patients being evaluated for TBM should have CSF tested for bacteria with Gram stain and culture and for fungi with India Ink preparation and cryptococcal antigen lateral flow assay [15]. *Cryptococcus neoformans* is the most common etiology of

Table 4.1 Microscopic and biochemical findings of cerebrospinal fluid in confirmed meningitis cases.[a]

Pathogen	References	*N*	White blood cells (per mm^3)	Lymphocytes (%)	Glucose (mmol/L)	Protein (g/L)
Virus	[11]	218	257 ± 520	71	3.4 ± 1.0	1.0 ± 0.6
Bacteria	[11]	35	1515 ± 2000	34	2.5 ± 1.5	4.9 ± 4.6
Cryptococcus (HIV-positive patients)	[12]	224	20 (5–60)	N.P.	2.3 (1.5–2.9)	N.P.
Mycobacterium tuberculosis (HIV-uninfected patients)	[b]	11	50 (25–100)	85 (80–90)	0.8 (0.3–1.0)	1.4 (1.0–2.0)
Mycobacterium tuberculosis (HIV-infected patients)	[13]	22	12 (3–140)	70 (65–85)	2.5 (1.5–3.8)	2.8 (1.1–5.5)

[a] Values given as mean ± standard deviation or median (interquartile range); *N.P.*, not provided.
[b] Author, personal observations, 2016–19.

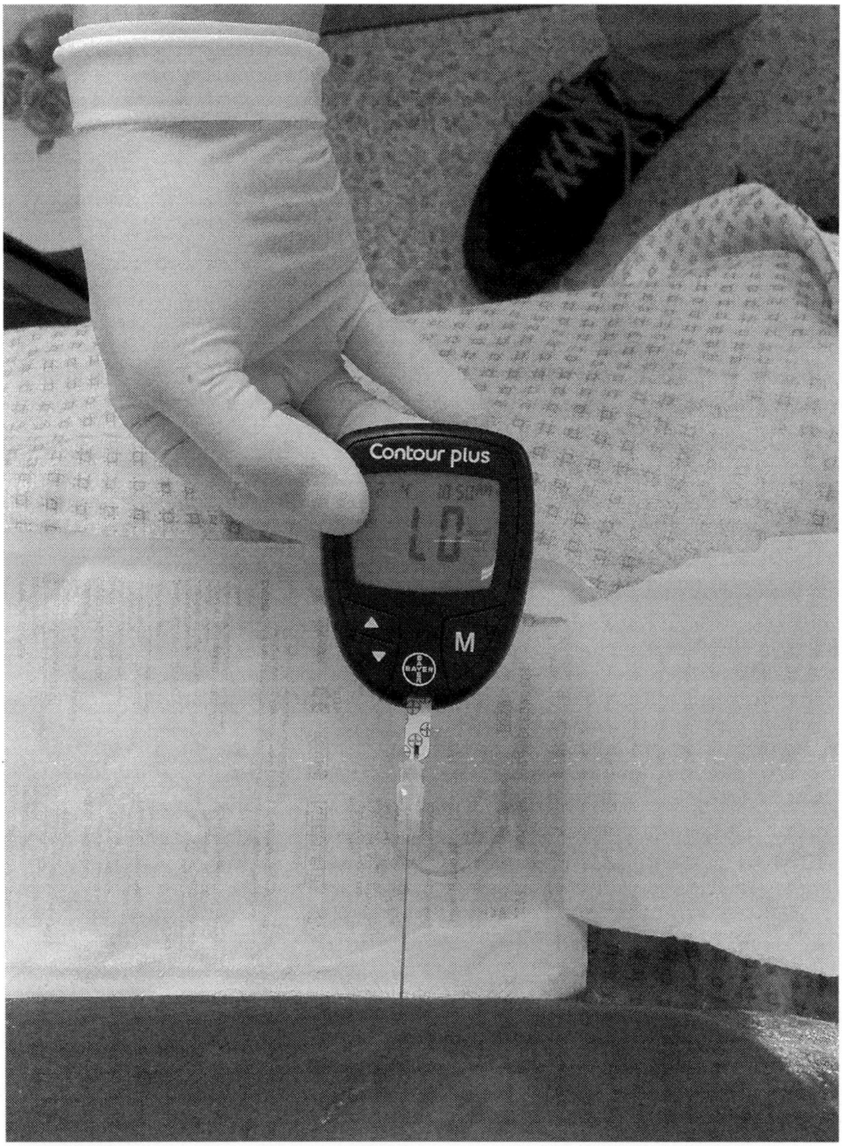

Figure 4.2 Direct measurement of CSF glucose (LO reading = less than 0.6 mmol/L). *CSF*, Cerebrospinal fluid.

meningitis in highly immunosuppressed individuals with HIV infection and has been reported as a cause of meningitis in immunocompetent individuals without HIV infection [16]. In addition, HIV-infected patients have been reported with concomitant TBM and cryptococcal meningitis [17].

Direct microscopy for diagnosis of tuberculosis

Examination of stained sputum specimens by microscopy is the most widely employed method for diagnosing pulmonary tuberculosis worldwide. The two methods recommended by the WHO are (1) conventional light microscopy of Ziehl–Neelsen (ZN) stained direct smears and (2) light-emitting diode fluorescence microscopy of auramine-stained smears [18]. Both of these methods have a moderate to high sensitivity against sputum culture as the reference standard [18]. In contrast, a multicenter study of conventional ZN staining of centrifuged CSF in cases of possible, probable, or definite TBM (clinical diagnosis as the gold standard) reported a low sensitivity of 33.9% [19]. The sensitivity improved to 66.4% using positive culture as the reference. Microscopy cannot distinguish MTB complex from nontuberculous *Mycobacteria* (e.g., *M. abscessus* complex, *M. avium* complex, and *M. fortuitum* complex).

Commercial nucleic acid amplification tests for *Mycobacterium tuberculosis* detection

The development of nucleic acid amplification tests (NAAT) for the detection of pathogens has simplified the rapid diagnosis of infectious diseases. Many hospital-based laboratories have developed in-house methods for detecting MTB DNA in clinical samples, including CSF. Tables 4.2 and 4.3 list the commercial testing platforms currently approved by the US Food and Drug Administration and/or the WHO for the identification of MTB and the determination of drug-resistance to first-line or second-line drugs. Xpert MTB/Rifampicin (RIF) (Cepheid, United States) is the most commonly used rapid diagnostic test used worldwide due to the preferential pricing for eligible countries following the endorsement by the WHO in 2010 [20]. Although Xpert MTB/RIF was approved for use in the United States in 2013 for testing sputum specimens, the WHO recommends Xpert MTB/RIF rather than conventional

Table 4.2 Nucleic acid–based tests for *Mycobacterium tuberculosis* approved by the US Food and Drug Administration.[a]

Trade name	Manufacturer
Xpert MTB/RIF assay	Cepheid
BDProbetec ET *M. tuberculosis* complex culture identification kit	Becton, Dickinson and Co.
Amplified *M. tuberculosis* direct test	Gen-Probe, Inc.
Amplicor *M. tuberculosis* test	Roche Molecular Systems, Inc.
SNAP *M. tuberculosis* complex	Syngene, Inc.
Accuprobe *M. tuberculosis* complex test	Gen-Probe, Inc.
Rapid Diagnostic System for *M. tuberculosis*	Gen-Probe, Inc.
Rapid Identification Test for *M. tuberculosis* complex	Gen-Probe, Inc.

[a]https://www.fda.gov/medical-devices/vitro-diagnostics/nucleic-acid-based-tests.

Table 4.3 Nucleic acid–based tests for *Mycobacterium tuberculosis* recommended by the World Health Organization.[a]

Trade name	Manufacturer
Xpert MTB/RIF assay	Cepheid
Xpert MTB/RIF Ultra assay	Cepheid
TB-LAMP	Eiken Chemical Company
GenoType MTBDR*plus* and Genotype MTBDR*sl*	Hain Lifescience
NTM + MDRTB detection kit 2	Nipro

[a]https://www.who.int/medical_devices/publications/Standalone_document_v8.pdf?ua = 1.

microscopy and culture as the first diagnostic test for suspected pulmonary and extrapulmonary tuberculosis, including TBM [21]. Xpert MTB/RIF is a fully automated cartridge-based technology requiring minimal technical training and can be performed under the same biosafety precautions as for smear microscopy (Fig. 4.3).

Xpert MTB/RIF requires a 2.0 mL sample of CSF that can be applied directly to the cartridge without dilution with the manufacturer-supplied sample reagent that is provided only for digestion and decontamination of sputum and tissue samples [8]. The test involves ultrasonic lysis of the sample to disrupt the MTB cell wall followed by reverse transcriptase—PCR amplification of an 81 base pair core region of the MTB complex

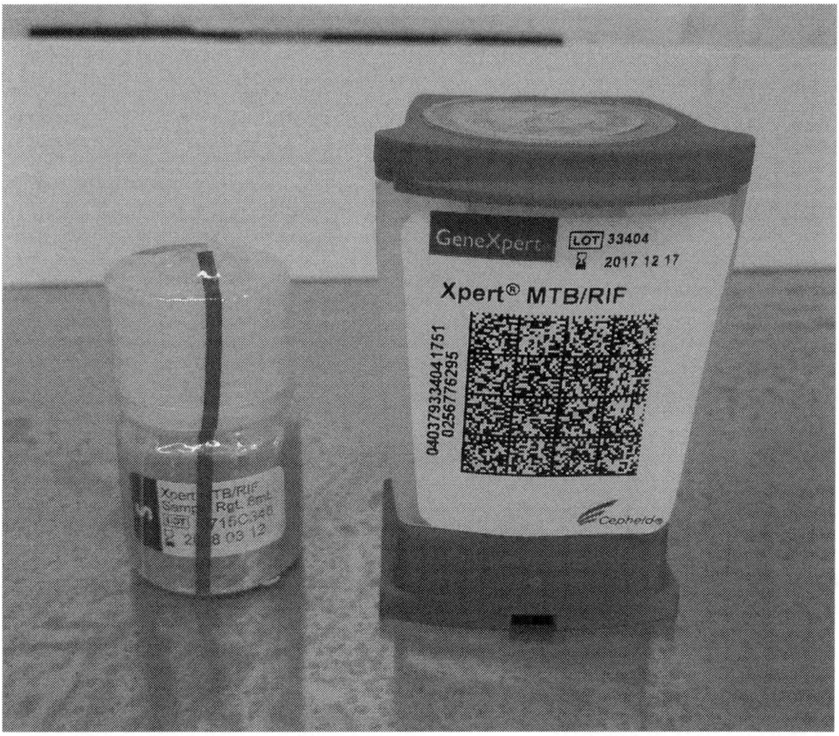

Figure 4.3 Xpert MTB/RIF: sample reagent (left) and single-use assay cartridge (right). *MTB, Mycobacterium tuberculosis. RIF,* Rifampicin.

specific rpoB gene (RNA polymerase Beta-subunit gene) that confers rifampicin (RIF) resistance. The rpoB sequence is probed with five fluorophore-labeled molecular beacons (Probes A-E). A sample processing control is included. Results are available within 2 hours and are reported as (1) MTB detected or not detected (semiquantitative categories of high, medium, low, and very low) and (2) RIF resistance detected, not detected, or indeterminate.

Although the sensitivity of Xpert MTB/RIF for detecting MTB in sputum specimens is high, the reported sensitivities in cases of suspected TBM is low [19,22]. The reported sensitivities vary depending on the gold standard used as the reference that may be a clinical diagnosis based on a case definition for TBM [23] and/or a composite of positive microbiological tests, including smear microscopy, Xpert MTB/RIF, other NAAT, and liquid or solid culture. Furthermore, the CSF processing protocols have varied widely between studies with differences in the use of sample reagent for dilution and the use of centrifugation [8]. The recently approved next-generation Xpert MTB/RIF Ultra (Ultra) appears to have an improved sensitivity for detecting MTB in CSF over Xpert MTB/RIF [13,24–26]. Ultra has three hybridization targets (rpoB, IS*6110*, IS*1081*) and a larger PCR reaction chamber. Bahr et al. reported sensitivities of 70% for Ultra, 43% for Xpert MTB/RIF, and 43% for liquid culture in 23 HIV-infected patients with probable or definite TBM [13]. Chin et al. reported sensitivities of 64% for Ultra, 36% for Xpert MTB/RIF, and 45% for liquid culture in 11 patients with probable or definite TBM (2/11 with HIV infection) [26].

A recent systematic review and metaanalysis of in-house and commercial NAAT for TBM reported pooled estimates of sensitivities of 82% against culture and 68% against a combined reference standard (defined as patients who fulfilled clinical criteria for TBM and had one or more positive results on culture, NAAT, or smear microscopy) [27]. Specificities were 99% against culture and 98% against combined reference standard. The following commercial tests were included in this review: Xpert MTB/RIF, Xpert MTB/RIF Ultra, Cobas Amplicor MTB, Gen-Probe MTD, Probe TEC, Abbott LCx ligase chain reaction, and GenoType MTBDRplus. The authors state, "Thus, our results suggest that a negative commercial NAA test result should not be used alone as a justification to rule out TBM." The reader is referred to the websites of the manufacturers of the commercial NAAT products included in this metaanalysis and in Tables 4.2 and 4.3 for technical information.

Adenosine deaminase test and lipoarabinomannan lateral flow assay

Adenosine deaminase (ADA) is an enzyme that converts adenosine to inosine and is present at high levels in activated T-lymphocytes. Measurement of ADA has been employed as a biomarker to support a diagnosis of TBM, but the clinical utility of this test is controversial because of a lack of standardization of assays and different cutoffs employed to distinguish nonmycobacterial bacterial meningitis from TBM [28–30]. A lateral flow assay to detect lipoarabinomannan, a cell wall component of mycobacteria, in urine (Alere Determine TB LAM Ag, Alere Inc, Waltham, MA, United States) was approved by the WHO in

2015 for the diagnosis of tuberculosis in HIV-infected adults with CD4 counts less than 100/μL or who are seriously ill regardless of CD4 count or with unknown CD4 count [31].

Next-generation sequencing

Next-generation sequencing (NGS) technologies are attracting interest worldwide for the improved and early diagnosis of infectious diseases, especially neuro-infectious diseases. NGS involves sequencing multiple DNA or RNA molecules in parallel in a single run and generating millions of reads of variable base pair lengths. NGS can be applied to cultured pathogens (whole-genome sequencing), enriched specimens using PCR amplification (targeted NGS), or to a specimen directly [metagenomic NGS (mNGS), unbiased NGS, agnostic NGS]. mNGS involves a complex bioinformatics process to compare the generated reads against databases of the human genome and pathogen genomes. In the United States, most laboratories performing mNGS for the diagnosis of infectious diseases offer testing on a research basis only. However, other countries, for example, China, already have a flourishing commercial industry for mNGS. There are numerous technical and quality-assurance requirements for reliable mNGS that are beyond the scope of this chapter. The reader is referred to excellent recent reviews [32,33]. At the time of the writing of this chapter, few studies have attempted to validate mNGS for the detection of MTB [34].

Immunological assays

The determination of cellular immune responses to MTB has been used to evaluate individuals for suspected latent MTB infection [35]. The traditional and least expensive method is the tuberculin skin test (TST) that is performed by administering an intradermal injection of purified-protein derivative (PPD) and measuring the transverse diameter of skin induration after 48–72 hours. Bacille Calmette–Guerin vaccination reduces the specificity of TST. Two PPD solutions are approved by the US Food and Drug Administration: Tubersol (Sanofi Pasteur Limited) and Aplisol (JHP Pharmaceuticals, LLC). Assays to measure interferon-gamma release (IGRA) from T-lymphocytes in response to MTB antigens have replaced TST in many countries. The currently available IGRA commercial tests are the QuantiFERON-TB Gold In-Tube and the QuantiFERON-TB

Gold Plus (Qiagen, The Netherlands) and the T-SPOT.TB (Oxford Immunotec, United Kingdom). Neither the TST nor IGRA tests should be used to rule in or rule out active tuberculosis, including TBM.

References

[1] WHO. Laboratory methods for the diagnosis of meningitis caused by *Neisseria meningitidis*, *Streptococcus pneumoniae*, and *Haemophilus influenzae*: WHO manual. 2nd ed. Geneva, Switzerland: World Health Organization; 2011. <https://apps.who.int/iris/handle/10665/70765> [accessed 21.07.19].

[2] McGill F, Heyderman RS, Panagiotou S, Tunkel AR, Solomon T. Acute bacterial meningitis in adults. Lancet 2016;388:3036−47.

[3] McGill F, Griffiths MJ, Bonnett LJ, Geretti AM, Michael BD, Beeching NJ, et al. Incidence, aetiology, and sequelae of viral meningitis in UK adults: a multicentre prospective observational cohort study. Lancet Infect Dis 2018;18:992−1003 Published online. Available from: https://doi.org/10.1016/S1473-3099(18)30245-7.

[4] Leber AL, Everhart K, Balada-Llasat J-M, Cullison J, Daly J, Holt S, et al. Multicenter evaluation of BioFire FilmArray Meningitis/Encephalitis Panel for detection of bacteria, viruses, and yeast in cerebrospinal fluid specimens. J Clin Microbiol 2016;54:2251−61.

[5] Liesman RM, Strasburg AP, Heitman AK, Theel ES, Patel R, Binnicker MJ. Evaluation of a commercial multiplex molecular panel for diagnosis of infectious meningitis and encephalitis. J Clin Microbiol 2018;56 e01927-17.

[6] North RJ, Izzo AA. Mycobacterial virulence. Virulent strains of *Mycobacteria tuberculosis* have faster in vivo doubling times and are better equipped to resist growth-inhibiting functions of macrophages in the presence and absence of specific immunity. J Exp Med 1993;177:1723−33.

[7] WHO. Tuberculosis laboratory biosafety manual. Geneva, Switzerland: World Health Organization; 2012. <http://www.who.int/tb/publications/2012/tb_biosafety/en/> [accessed 21.07.19].

[8] Chin JH, Ssengooba W, Grossman S, Pellinen J, Wadda V. Xpert MTB/RIF Ultra: optimal procedures for the detection of *Mycobacterium tuberculosis* in cerebrospinal fluid. J Clin Tuberc Other Mycobact Dis 2019;14:16−18.

[9] Rousseau G, Asmolov R, Grammatico-Guillon L, Auvet A, Laribi S, Garot D, et al. Rapid detection of bacterial meningitis using a point-of-care glucometer. Eur J Emerg Med 2019;26(1):41−6.

[10] Mlinarić A, Vogrinc Z, Drenšek Z. Effect of sample processing and time delay on cell count and chemistry tests in cerebrospinal fluid collected from drainage systems. Biochem Med (Zagreb) 2018;28(3):030705.

[11] Viallon A, Desseigne N, Marjollet O, Birynczyk A, Belin M, Guyomarch S, et al. Meningitis in adult patients with a negative direct cerebrospinal fluid examination: value of cytochemical markers for differential diagnosis. Crit Care 2011;15:R136.

[12] Beardsley J, Wolbers M, Kibengo FM, Ggayi A-BM, Kamali A, Cuc NTK, et al. Adjunctive dexamethasone in HIV-associated cryptococcal meningitis. N Engl J Med 2016;374:542−54.

[13] Bahr NC, Nuwagira E, Evans EE, Cresswell FV, Bystrom PV, Byamukama A, et al. Diagnostic accuracy of Xpert MTB/RIF Ultra for tuberculous meningitis in HIV-infected adults: a prospective cohort study. Lancet Infect Dis 2018;18:68−75.

[14] Cresswell FV, Bangdiwala A, Meya DB, Vidal JE, Török ME, Thao LTP, et al. Absence of cerebrospinal fluid pleocytosis in tuberculous meningitis is a common occurrence in HIV co-infection and a predictor of poor outcomes. Int J Infect Dis 2018;68:77−8.

[15] Lourens A, Jarvis JN, Meintjes G, Samuel CM. Rapid diagnosis of cryptococcal meningitis by use of lateral flow assay on cerebrospinal fluid samples: influence of the high-dose "hook" effect. J Clin Microbiol 2014;52:4172–5.
[16] Williamson PR, Jarvis JN, Panackal AA, Fisher MC, Molloy SF, Loyse A, et al. Cryptococcal meningitis: epidemiology, immunology, diagnosis and therapy. Nat Rev Neurol 2017;13:13–24.
[17] Ellis J, Cresswell FV, Rhein J, Ssebambulidde K, Boulware DR. Cryptococcal meningitis and tuberculous meningitis co-infection in HIV-infected Ugandan adults. Open Forum Infect Dis 2018;5:ofy193. Available from: https://doi.org/10.1093/ofid/ofy193.
[18] WHO. Fluorescent light-emitting diode (LED) microscopy for diagnosis of tuberculosis. Geneva, Switzerland: World Health Organization; 2011. <https://www.who.int/tb/publications/2011/led_microscopy_diagnosis_9789241501613/en/> [accessed 21.07.19].
[19] Heemskerk AD, Donovan J, Thu DDA, Marais S, Chaidir L, Dung VTM, et al. Improving the microbiological diagnosis of tuberculous meningitis: a prospective, international, multicentre comparison of conventional and modified Ziehl–Neelsen stain, GeneXpert, and culture of cerebrospinal fluid. J Infect 2018;77:509–15. Available from: https://doi.org/10.1016/j.jinf.2018.09.003.
[20] Albert H, Nathavitharana RR, Isaacs C, Pai M, Denkinger CM, Boehme CC. Development, roll-out and impact of Xpert MTB/RIF for tuberculosis: what lessons have we learnt and how can we do better? Eur Respir J 2016;48:516–25.
[21] WHO. Xpert MTB/RIF implementation manual: technical and operational 'how-to'; practical considerations. Geneva, Switzerland: World Health Organization; 2014. <http://www.who.int/tb/publications/xpert_implem_manual/en/> [accessed 21.07.19].
[22] Kohli M, Schiller I, Dendukuri N, Dheda K, Denkinger CM, Schumacher SG, et al. Xpert® MTB/RIF assay for extrapulmonary tuberculosis and rifampicin resistance. Cochrane Database Syst Rev 2018;8. Available from: https://doi.org/10.1002/14651858.CD012768.pub2 Art. No.: CD012768.
[23] Marais S, Thwaites G, Schoeman JF, Török ME, Misra UK, Prasad K, et al. Tuberculous meningitis: a uniform case definition for use in clinical research. Lancet Infect Dis 2010;10:803–12.
[24] WHO. WHO meeting report of a technical expert consultation: non-inferiority analysis of Xpert MTB/RIF Ultra compared to Xpert MTB/RIF. Geneva, Switzerland: World Health Organization; 2017. <http://www.who.int/tb/publications/2017/XpertUltra/en/> [accessed 21.07.19].
[25] Chakravorty S, Simmons AM, Rowneki M, Cao Y, Ryan J, Banada PP, et al. The new Xpert MTB/RIF Ultra: improving detection of *Mycobacterium tuberculosis* and resistance to rifampin in an assay suitable for point-of-care testing. mBio 2017;8:e00812–17.
[26] Chin JH. Xpert MTB/RIF Ultra: the long-awaited game changer for tuberculous meningitis? Eur Respir J 2017;50(4):1701201. Available from: https://doi.org/10.1183/13993003.01201-2017.
[27] Pormohammad A, Nasiri MJ, McHugh TD, Riahi SM, Bahr NC. A systematic review and meta-analysis of the diagnostic accuracy of nucleic acid amplification tests for tuberculous meningitis. J Clin Microbiol 2019;57:e01113–18.
[28] Tuon FF, Higashino HR, Lopes MIBF, Litvoc MN, Atomiya AN, Antonangelo L, et al. Adenosine deaminase and tuberculous meningitis—a systematic review with meta-analysis. Scand J Infect Dis 2010;42:198–207.

[29] Ekermans P, Dusé A, George J. The dubious value of cerebrospinal fluid adenosine deaminase measurement for the diagnosis of tuberculous meningitis. BMC Infect Dis 2017;17:104. Available from: https://doi.org/10.1186/s12879-017-2221-3.
[30] Pormohammad A, Riahi S-M, Javad Nasiri MJ, Fallah F, Aghazadeh M, Doustdar F, et al. Diagnostic test accuracy of adenosine deaminase for tuberculous meningitis: a systematic review and meta-analysis. J Infect 2017;74:545–54.
[31] WHO. The use of lateral flow urine lipoarabinomannan assay (LF-LAM) for the diagnosis and screening of active tuberculosis in people living with HIV. Geneva, Switzerland: World Health Organization; 2015. <https://www.who.int/tb/publications/use-of-lf-lam-tb-hiv/en/> [accessed 21.07.19].
[32] Simner PJ, Miller S, Carroll KC. Understanding the promises and hurdles of metagenomic next-generation sequencing as a diagnostic tool for infectious diseases. Clin Infect Dis 2018;66:778–88.
[33] Boers SA, Jansen R, Hays JP. Understanding and overcoming the pitfalls and biases of next-generation sequencing (NGS) methods for use in the routine clinical microbiological diagnostic laboratory. Eur J Clin Microbiol Infect Dis 2019;38:1059–70.
[34] Miao Q, Ma Y, Wang Q, Pan J, Zhang Y, et al. Microbiological diagnostic performance of metagenomic next-generation sequencing when applied to clinical practice. Clin Infect Dis 2018;67:S231–40.
[35] Pai M, Denkinger CM, Kik SV, Rangaka MX, Zwerling A, Oxlade O, et al. Gamma Interferon release assays for detection of *Mycobacterium tuberculosis* infection. Clin Microbiol Rev 2014;27:3–20.

CHAPTER 5

Identification of *Mycobacterium tuberculosis* drug resistance

Willy Ssengooba[1] and Jerome H. Chin[2]
[1]Mycobacteriology Unit, Department of Medical Microbiology, Makerere University, Kampala, Uganda
[2]Department of Neurology, NYU Langone Health, New York, NY, United States

Key points

- The burden of tuberculosis (TB) and drug-resistant TB remains high despite the enormous efforts towards TB control.
- There are known and emerging mechanisms of drug resistance to both existing and emerging anti-TB drugs.
- Culture-based and molecular drug-susceptibility testing methods are available with differing strengths and limitations.

Global burden of tuberculosis and drug resistance

Human tuberculosis (TB) is a disease of global public health importance declared in 2015 as the deadliest infectious disease alongside HIV/AIDS. As such, the World Health Organization's (WHO) End TB Strategy calls for a 95% reduction in TB deaths and a 90% reduction in TB incidence rate by 2035 compared with 2015 [1]. According to the 2018 WHO global report for TB [2], there were an estimated 10 million incident cases of TB worldwide in 2017, of which 90% were adults and 9% were people living with HIV/AIDS (PLWH). In 2017, TB caused an estimated 1.3 million deaths among HIV-negative people and an additional 300,000 deaths among PLWH. There were an estimated 558,000 people (range, 483,000−639,000) with resistance to rifampicin (RR-TB), of which 82% had multidrug-resistant TB (MDR-TB), that is, TB that is resistant to both rifampicin and isoniazid, the two most important drugs used in the first-line regimen to treat uncomplicated TB. Extensively drug-resistant TB, TB resistant to isoniazid and rifampicin plus one of the fluoroquinolones and at least one of the three second-line injectable drugs, accounted for 8.5% of the total cases of MDR-TB. There are still huge gaps between diagnosis and treatment initiation for MDR-TB.

Tuberculous Meningitis
DOI: https://doi.org/10.1016/B978-0-12-818825-5.00005-X

TB is not only disproportionately present in certain populations but also in certain regions around the globe. The WHO African and Southeast Asia regions have the highest incidence rates [2]. Collectively, the global rate of decline in TB incidence of only 2.0% is too slow to permit reaching the goals of the WHO End TB Strategy. Universal screening for drug resistance, TB treatment informed by drug resistance patterns, and use of shorter regimens with drugs that are more effective are some of the key actions recommended by the WHO End TB Strategy toward reduction of the MDR-TB burden. Rapid detection of drug-resistant TB is one of the critical elements for improved TB control. TB patients at higher risk of MDR-TB should be tested for individual and public health benefits. The most affected countries with MDR-TB cases are mainly challenged with insufficient laboratory capacity to aid early diagnosis, leading to inappropriate patient management.

Mechanisms of resistance to antituberculosis drugs

Resistance to antituberculosis (anti-TB) drugs may be through genetic or phenotypic changes. The strategies, natural and acquired, used by the bacteria may include the following: modification of the drug targets (e.g., *rpoB* and rifampicin resistance), modification of drug activation or metabolic pathways (e.g., *katG* and isoniazid resistance, *pncA* and pyrazinamide resistance), modification of drug target amplification (e.g., *inhA* and isoniazid resistance), hydrolytic or drug-modifying enzymes (e.g., β-lactamases and aminoglycoside acetyltransferases), and barrier mechanisms (decreased permeability and efflux pumps). The last two mechanisms are the main drivers of the natural resistance of *Mycobacteria* to most of the anti-TB drugs. Table 5.1 lists the mechanisms and genes associated with resistance to anti-TB drugs, adapted from Hatfull et al. [3] and Mathema et al. [4].

Drug resistance determination methods: strengths and limitations

Results of drug-susceptibility testing (DST) are vital for the proper management of TB patients, especially in areas with a high burden of DR-TB, and are a key component of the WHO End TB Strategy. Results of DST are not only useful for individualized patient management but also in clinical trials for eligibility, arm allocation, and end point determination. There are two main forms of DST: (1) culture-based DST and

Table 5.1 Mechanisms and genes associated with drug resistance to antituberculosis drugs.

Drug	Mechanism of action	Gene(s) involved in resistance	Role	Product	Mutation frequency (%) among resistant *Mycobacterium tuberculosis*
Streptomycin	Inhibition of protein synthesis	*rpsL*	Drug target	Ribosomal protein S12	52–59
		rrs	Drug target	16S rRNA	8–21
Isoniazid	Inhibition of mycolic acid biosynthesis and other potential multiple effects on DNA, lipid, carbohydrates, and NAD metabolism	*katG*	Conversion of prodrug	Catalase–peroxidase	42–58
		inhA	Drug target	Enoyl-ACP reductase	21–34
		kasA	Drug target	β-Ketoacyl-ACP synthase	<10
		oxyR-ahpC	Marker of resistance	Alkyihydroreductase	10–15
Rifampicin	Inhibition of transcription	*rpoB*	Drug target	β subunit RNA polymerase	96–100
Ethambutol	Inhibition of arabinogalactan synthesis	*embCAB*	Drug target	Arabinosyltransferases	47–65
Pyrazinamide	Acidification of cytoplasm	*pncA*	Conversion of prodrug	Amidase	72–97

(*Continued*)

Table 5.1 (Continued)

Drug	Mechanism of action	Gene(s) involved in resistance	Role	Product	Mutation frequency (%) among resistant *Mycobacterium tuberculosis*
Amikacin/kanamycin	Inhibition of protein synthesis	*rrs*	Drug target	16S rRNA	76
Fluoroquinolones	Inhibition of DNA gyrase	*gyrA*/*gyrB*	Drug target	DNA gyrase α/β subunits	75–94[a]
Ethionamide	Inhibition of mycolic acid biosynthesis	*inhA*/*ethA*	Drug target	Enoyl-ACP reductase/flavoprotein monooxygenase	<10[b]

[a] *gyrA* only.
[b] *inhA* only.

(2) molecular DST. Culture-based methods are used for diagnosis and DST, and both components are discussed in this chapter. Nucleic acid amplification tests (NAAT) for the diagnosis of TB and TBM are reviewed in Chapter 4, Laboratory methods for detecting tuberculosis and tuberculous meningitis. Here, we elaborate on DNA-based assays for DST.

Culture-based drug-susceptibility testing for *Mycobacterium tuberculosis*

Culture procedures for TB diagnosis start with sample processing to eliminate other floras that are considered contaminants. For decontamination, sodium hydroxide, sodium citrate, and *N*-acetyl-L-cysteine are added to a sputum sample to a final concentration of 1%–1.5% of sodium hydroxide. Nonblood-containing sterile samples such as cerebrospinal fluid (CSF) are inoculated directly into culture media without this decontamination process. After 15 minutes of room temperature decontamination, the mixture is diluted with phosphate-buffered saline (PBS; pH 6.8) to neutralize the decontaminating solution. The diluted sample is then centrifuged at $3000 \times g$ for 20 minutes, and the supernatant is decanted. The pellet is reconstituted in 2 mL of PBS and now ready for inoculation. Three drops of concentrated sample are added onto solid culture media, including Löwenstein–Jensen medium (LJ; up to 8 weeks for a negative) and Middlebrook 7H10/7H11 (MB7H10/11; up to 3 weeks for a negative), or into broth media, including Middlebrook 7H9 [mycobacterial growth indicator tube (MGIT; Becton, Dickinson, and Company, Sparks, Maryland, United States) and microscopic observation of drug susceptibility (MODS)].

Culture in MGIT is done according to the manufacturer's manual [5]. Briefly, a growth supplement together with a mixture of antibiotics [polymyxin B, amphotericin B, nalidixic acid, trimethoprim, and azlocillin (PANTA)] is added to the MGIT tubes before inoculation. A volume of 0.5 mL of sample is inoculated into MGIT and then incubated in the MGIT machine which automatically monitors for growth up to 6 weeks for a negative. The addition of PANTA to CSF specimens is not necessary and up to 0.8 mL of CSF can be inoculated. Because the growth is usually not specific, a Ziehl–Neelsen stained smear is done to confirm mycobacterial growth and a rapid identification test for *M. tuberculosis* complex (MTB complex) using MPT64 antigen kit is done. Only pure cultures

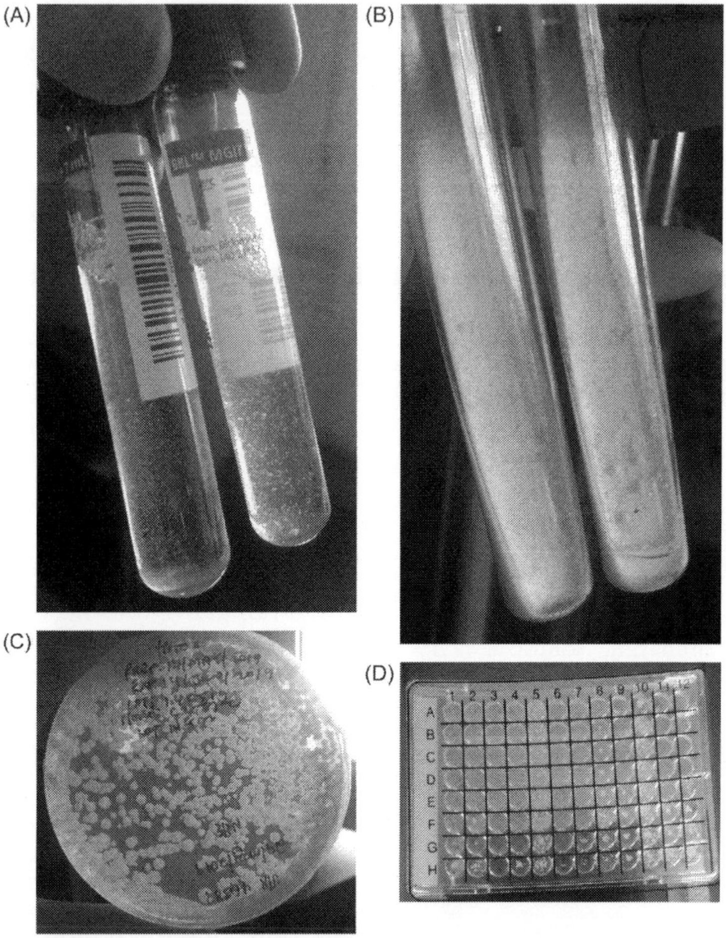

Figure 5.1 ***Different media used for*** **Mycobacteria** ***culture and drugs susceptibility testing.*** (A) MGIT, (B) Löwenstein–Jensen medium, (C) Middlebrooke 7H10 medium, and (D) Sensititre MYCOTB plate. *MGIT*, Mycobacterial growth indicator tube.

with no growth on blood agar culture after 24 hours are considered for DST (Fig. 5.1).

Mycobacterial growth indicator tube drug-susceptibility testing

The MGIT-DST is a liquid culture form of DST using the automated BACTEC MGIT 960 System (Becton, Dickinson, and Company, Sparks, Maryland, United States). For safety concerns, it is recommended that this procedure is carried out in a biosafety level three (BSL-3) facility within a class II biosafety cabinet. The MGIT-DST has been validated for all WHO-approved first-line drugs and a number of second-line drugs used in TB treatment. Although the MGIT-DST method is endorsed by WHO for both direct (using smear-positive samples) and indirect (using positive cultures) applications, the indirect DST is the most implemented method worldwide. The indirect MGIT-DST involves obtaining pure MTB growth and preparation of MGIT with standard critical concentrations of TB drugs or with no drug (control). A standardized inoculum of pure cultures is applied to the drug-containing and control tubes that are then incubated in the MGIT machine at 37°C until a standard growth is achieved for the test to be interpreted. When the growth unit (GU) of the control tubes reaches 400 within 4—13 days (streptomycin, isoniazid, rifampicin, ethambutol) or 4—21 days (pyrazinamide), the GU values of the drug-containing tubes are evaluated. A patient will be considered to have drug-susceptible TB if the GU of a drug-containing tube is less than 100; otherwise, the patient will be considered to have TB resistant to that specific TB drug. The concentrations approved for each drug according to the employed media are highlighted in the WHO critical concentrations technical report [6]. MGIT-DST is considered to be highly sensitive with a short turnaround time (TAT). However, the costs for the machines and reagents and requirement for a BSL-3 make this method too costly for many laboratories in low- and middle-income countries (LMICs).

Agar proportion drug-susceptibility testing methods (Löwenstein—Jensen medium and MB7H10/11)

For DST using MB7H10/11, anti-TB drugs are prepared in stock concentrations and stored frozen. The stock concentrations are thawed, used to prepare the final/critical concentrations, and discarded. Media are prepared in the same way as for initial MTB cultures. The stock drug solutions are diluted with distilled water to make the working solutions and then mixed with the media to achieve the desired critical drug concentration [6]. Standard culture suspensions are made at 10^{-2} and 10^{-4} dilutions followed by inoculation of 100 μL (three drops) onto each drug-containing quadrant as well as a plain plate as a control and streak out. The plates are then incubated between 35°C and 38°C and are interpreted at day 21. The 10^{-4} diluted suspension control is used to interpret an MTB culture as resistant or sensitive to a certain drug. The colony count of the 10^{-2} diluted suspension control should be higher than that of 10^{-4} diluted suspension control. If this is not the case, the DST is invalid. For a percentage resistance of 0 to <1% (99% growth inhibition), the drug is interpreted as sensitive. For a percentage resistance of ≥ 1%, the drug is interpreted as resistant.

DST using LJ is by the modified proportional method of Canetti [7]. Individual first-line and/or second-line drugs are added before inspissation (thickening of the media by evaporation of the fluid) [6]. The DSTs are set up with a standard inoculum prepared from the growth on primary diagnostic LJ or on a subculture using 10-fold serial dilutions. As a growth control, drug-free slopes are set up for each strain tested. After a time of incubation between 35°C and 38°C, the number of colonies growing on drug-free medium is compared with the number on drug-containing medium and the proportion of resistant organism is calculated. Resistance is determined by the number of colonies on drug-containing media (at the critical concentrations of the drugs with the 10^{-2} culture suspension dilution) compared to the growth on drug-free medium inoculated with the 10^{-4} culture suspension dilution. If the number of colonies on the drug-containing medium is greater than control drug-free medium, a strain is considered to be resistant to that drug; otherwise, it is considered to be susceptible. The agar proportion methods of drug resistance testing, although less expensive than MGIT-DST, are more complex and have a longer TAT (up to 6 weeks). In addition, these methods require high biosafety and technical standards and are prone to contamination.

Sensititre Mycobacterium tuberculosis MYCOTB AST plate

The Sensititre Mycobacterium tuberculosis MYCOTB AST plate (Thermo Fisher Scientific, Massachusetts, United States) for susceptibility testing of MTB contains 12 lyophilized TB drugs, including first-line and second-line drugs, in the plate wells [8]. It is an indirect method that requires pure culture colonies that are emulsified into saline and 0.2% Tween and adjusted to a turbidity of 0.5 McFarland standard. This is further adjusted to an inoculum of 1×10^5 cfu/mL that is later poured into a trough and inoculated into the plate using a multichannel pipette. Fig. 5.2 shows the schema for the plate. The plate is then sealed and incubated at 37°C and examined visually for signs of contamination 48−72 hours after inoculation followed by days 7, 10, 14, and 21 readings. Interpretation of results is only possible when adequate growth is obtained in the positive control wells. Minimum inhibitory concentration is recorded as the lowest antibiotic concentration that reduces visible growth. This method requires

Wells	1	2	3	4	5	6	7	8	9	10	11	12
A	OFL 32	MOX 8	RIF 16	AMI 16	STR 32	RFB 16	PAS 64	ETH 40	CYC 256	INH 4	KAN 40	EMB 32
B	OFL 16	**MXF 4**	RIF 8	**AMI 8**	**STR 16**	RFB 8	PAS 32	ETH 20	CYC 128	**INH 2**	KAN 20	EMB **16**
C	OFL 8	**MOX 2**	RIF 4	**AMI 4**	**STR 8**	RFB 4	PAS 16	**ETH 10**	CYC 64	**INH 1**	**KAN 10**	**EMB 8**
D	**OFL 4**	**MOX 1**	**RIF 2**	**AMI 2**	**STR 4**	RFB 2	PAS 8	**ETH 5**	**CYC 32**	**INH 0.5**	**KAN 5**	**EMB 4**
E	**OFL 2**	**MOX 0.5**	**RIF 1**	AMI 1	**STR 2**	**RFB 1**	**PAS 4**	**ETH 2.5**	**CYC 16**	**INH 0.25**	**KAN 2.5**	EMB 2
F	**OFL 1**	**MOX 0.25**	**RIF 0.5**	AMI 0.5	**STR 1**	**RFB 0.5**	**PAS 2**	ETH 1.2	CYC 8	**INH 0.12**	KAN 1.2	EMB 1
G	OFL 0.5	MOX 0.12	RIF 0.25	AMI 0.25	STR 0.5	**RFB 0.25**	**PAS 1**	ETH 0.6	CYC 4	INH 0.06	KAN 0.6	EMB 0.5
H	OFL 0.25	MOX 0.06	RIF 0.12	AMI 0.12	STR 0.25	RFB 0.12	PAS 0.5	ETH 0.3	CYC 2	INH 0.03	**POS**	**POS**

Figure 5.2 ***Schema for the Sensititre MYCOTB plate with drugs and critical concentrations.*** Shaded boxes indicate range for minimum inhibitory concentrations to be considered in agreement with APM. APM critical concentrations (μg/mL); OFL (2.0), MOX (0.5 low, 2.0 high), RIF (1.0), AMI (4.0), STR (2.0 low, 10.0 high), RFB (0.5), PAS (2.0), ETH (5.0), CYC (25), INH (0.2 low, 1.0 high), KAN (5.0), EMB (5.0 low, 10.0 high). *AMI*, Amikacin; *APM*, agar proportion methods; *CYC*, cycloserine; *EMB*, ethambutol; *ETH*, ethionamide; *INH*, isoniazid; *KAN*, kanamycin; *MOX*, moxifloxacin; *OFL*, ofloxacin; *PAS*, *P*-aminosalicylic acid; *RFB*, rifabutin; *RIF*, rifampicin; *STR*, streptomycin.

purchase of plates from the manufacturer that has limited distribution channels in LMICs. The method requires pure cultures that increase the TAT although the majority of results after plate setup can be interpreted within 14 days [8].

Noncommercial methods

In 2010 WHO endorsed several direct noncommercial culture DST for rapid screening of patients at risk of MDR-TB that can be used in laboratories that lack access to more sophisticated infrastructure and techniques [9]. These tests include the micro-well plate MODS assay and the nitrate reductase assay (NRA) [10,11]. Both DST are approved to be used directly on smear-positive sputum specimens that permit the early detection of MDR-TB in patients at high risk. They enable laboratories to be independent of more expensive commercial tests and both can be performed in biosafety level 2 laboratories that are usually available at a country's regional-level laboratories. The MODS assay has a pooled estimate of 98% for sensitivity (confidence Interval (CI) 95%—99%) and a pooled estimate of 99% for specificity (CI 96%–100%) for the detection of rifampicin resistance and slightly lower pooled estimate of sensitivity for isoniazid (91%; CI 87%–95%) [9]. It has a time to positivity of 6.5–11.6 days for direct and indirect applications, respectively. The standardized drug concentrations provided with this method's kit makes it user-friendly and less prone to errors of making the correct drug concentrations. This method initially had safety issues around the sealing of the plate wells; however, the advanced version has a robust rubber seal that reduces this safety concern.

The NRA is based on LJ incorporated with potassium nitrate at a concentration of 1000 mg/mL into which TB drugs at standard concentrations are added. The method can be done with positive smear sputum samples. NRA was found to have a pooled estimate of 97% for sensitivity (CI 95%–98%) and a pooled estimate of 100% for specificity (CI 99%–100%) for the detection of rifampicin resistance. For the detection of isoniazid resistance, a pooled sensitivity of 97% (CI 95%–98%) and a pooled specificity of 99% (CI 99%–100%) were obtained [9]. It has a time to positivity of 9–21 days for direct and indirect applications, respectively, and is affordable in most LMICs. Both MODS and NRA have not been fully implemented in LMICs due to the safety requirements for culturing MTB.

DNA-based tuberculosis drug-susceptibility testing methods

The increasing burden of MDR-TB has spurred the development of molecular diagnostics for DR-TB over the past decade. They have the advantages of being rapid and not requiring the same special biosafety equipment and protocols as for culture-based DST methods. Here we discuss the WHO-endorsed molecular tests for the detection of DR-TB, including line probe assays (LPA) and NAAT.

Xpert MTB/RIF and Xpert MTB/RIF Ultra

Xpert MTB/RIF and Xpert MTB/RIF Ultra (Cepheid, California, United States) are fully automated NAAT that employ a single-use cartridge to detect MTB as well as RR within 2 hours (see Chapter 4: Laboratory methods for detecting tuberculosis and tuberculous meningitis, for technical details) [12–14]. Both tests run on the GeneXpert platform. In 2013 the WHO recommended the use of Xpert MTB/RIF as the initial diagnostic test for extrapulmonary TB, including tuberculous meningitis in adults and children [13]. The next-generation Xpert MTB/RIF Ultra was endorsed by the WHO in 2018 as an alternative for Xpert MTB/RIF [14]. Rifampicin resistance determination is rapid and inexpensive compared to culture methods. Since almost all MDR-TB isolates are resistant to rifampicin, detection of RR by Xpert MTB/RIF or Xpert MTB/RIF Ultra is taken as a proxy for additional resistance to isoniazid. The main limitations of the assay are the restricted coverage of the rifampicin resistance-determining region (81 base pair region) and the lack of susceptibility information on other first-line drugs. The test equipment does incur high repair and calibration costs.

Line probe assays

LPA were first endorsed by the WHO for the rapid detection of MDR-TB in 2008 and updated policies were provided in 2016 [15,16]. There are three approved LPA: Genotype MTBDR*plus* (Hain Lifescience, Nehren, Germany) for detection of RR and isoniazid, Genotype MTBDR*sl* (Hain Lifescience) for detection of resistance to injectable agents (including kanamycin, amikacin, and capreomycin) and fluoroquinolone drugs (including ofloxacin, levofloxacin, moxifloxacin, and gatifloxacin), and Nipro NTM + MDRTB Detection Kit 2 (Nipro, Tokyo, Japan) for detection of RR and isoniazid. These tests can be done directly on smear-positive sputum samples and deliver results between 2 and 4 days compared to culture-based DST that can take up to 8 weeks. LPA is a strip-based test that involves MTB DNA extraction either from culture isolates or directly from sputum samples and then amplification using polymerase chain reaction (PCR) in a thermocycler. The amplified product is hybridized with specific oligonucleotide probes immobilized on the test strip with stringent washing that leads to the development of colored bands. This process enables the detection of MTB as well as mutations bands within resistance-determining regions known to be causing resistance to first-line or second-line drugs. The downsides of LPA are the need for a three-room set up in the laboratory (pre-PCR, PCR, and post-PCR rooms) and the technical demands and experience required to perform the assays effectively.

Next-generation sequencing

Next-generation sequencing (NGS) technologies are transitioning from the bench to the clinic for the management of infectious diseases. NGS utilizes direct sequencing methods to reconstruct the complete genome of a microorganism [whole-genome sequencing (WGS)] or specific genomic regions (targeted NGS) using commercial platforms (e.g., Illumina, Thermo Fisher Scientific, Pacific Biosciences, Oxford Nanopore Sequencing Technologies) followed by mapping of the results (reads) to a reference genome to detect genomic variants. Applications of NGS include strain typing, surveillance, outbreak investigation, determination of transmission patterns, and drug resistance and sensitivity profiling. The latter is particularly useful for the treatment of TB since MTB is a slow-growing organism, and conventional phenotypic DST results take many weeks for results. For details on the technical and analytical aspects as well as the limitations of NGS, the reader is referred to recent publications [17–19].

Although WHO-recommended rapid genetic tests (NAAT and LPA) can identify common preselected drug resistance mutations in positive clinical specimens [20], many less common resistance mutations are not detected by these assays. Results of WGS are analyzed using bioinformatics to identify single nucleotide polymorphisms and insertions or deletions that are associated with resistance to first- and second-line anti-TB drugs. Most published studies of WGS for MTB DST have been conducted on culture isolates from sputum [21–23]. Excellent sensitivities ($>91\%$) and specificities ($>93\%$) have been reported for WGS using phenotypic DST as the gold standard for the determination of drug sensitivity and resistance profiles for first-line drugs (isoniazid, rifampicin, ethambutol, pyrazinamide) in over 10,000 isolates from 16 countries [21]. The application of WGS directly to clinical samples would eliminate the culture delay and allow rapid individualized treatment decisions. However, contamination by host and normal microflora reads creates challenges to the generation of sufficient data for drug-susceptibility predictions [24]. Targeted NGS methods have been developed that include multiple resistance-associated gene targets to detect DR-TB direct from sputum [25,26]. WGS and targeted NGS of CSF specimens from patients with TBM for DST have not yet been evaluated and limited studies of NGS for the detection of MTB in CSF for diagnostic purposes have been published (see Chapter 4: Laboratory methods for detecting tuberculosis and tuberculous meningitis).

The high reliability, rapid turnaround, and declining cost of WGS are positioning WGS to replace traditional phenotypic procedures for the identification of DR-TB. In 2018 both England (Public Health England) and the United States (Centers for Disease Control and Prevention) launched WGS for DST. Expanding WGS to LMICs will require major investments in sequencing platforms, bioinformatics training, and computer and data storage infrastructure. The WHO has published a technical guide for the implementation of WGS and targeted NGS for the detection of drug-resistant TB in reference laboratories in LMICs [18].

References

[1] WHO. The End TB Strategy. Geneva, Switzerland: World Health Organization. <http://www.who.int/tb/strategy/end-tb/en/> [accessed 11.07.19].

[2] WHO. Global tuberculosis report. Geneva, Switzerland: World Health Organization; 2018. <https://www.who.int/tb/publications/global_report/en/> [accessed 11.07.19].

[3] Hatfull GF. Molecular genetics of mycobacteriophages. In: Hatfull GF, Jacobs Jr. WR, editors. Molecular genetics of mycobacteria. Washington, DC: ASM Press; 2000. p. 236.

[4] Mathema B, Kurepina NE, Bifani PJ, Kreiswirth BN. Molecular epidemiology of tuberculosis: current insights. Clin Microbiol Rev 2006;19(4):658–85.

[5] Siddiqi SH, Rüsch-Gerdes S. MGIT procedure manual for BACTEC MGIT 960 TB system (also applicable for manual MGIT) mycobacteria growth indicator tube (MGIT) culture and drug susceptibility demonstration projects. <http://www.finddx.org/wp-content/uploads/2016/02/mgit_manual_nov2006.pdf>; 2006 [accessed 11.07.19].

[6] WHO. Technical report on critical concentrations for drug susceptibility testing of medicines used in the treatment of drug-resistant tuberculosis. Geneva, Switzerland: World Health Organization; 2018. <https://www.who.int/tb/publications/2018/WHO_technical_report_concentrations_TB_drug_susceptibility/en/> [accessed 11.07.19].

[7] Canetti G, Froman S, Grosset J, Hauduroy P, Langerova M, Mahler HT, et al. Mycobacteria: laboratory methods for testing drug sensitivity and resistance. Bull World Health Organ 1963;29:565–78.

[8] Lee J, Armstrong DT, Ssengooba W, Park JA, Yu Y, Mumbowa F, et al. Sensititre MYCOTB MIC plate for testing *Mycobacterium tuberculosis* susceptibility to first- and second-line drugs. Antimicrob Agents Chemother 2014;58(1):11–18.

[9] WHO. Non-commercial culture and drug-susceptibility testing methods for screening of patients at risk of multi-drug resistant tuberculosis. Geneva, Switzerland: World Health Organization; 2010. <https://www.who.int/tb/laboratory/whopolicy_noncommercialculture_and_dstmethods_july10_revnov10.pdf> [accessed 11.07.19].

[10] Solis LA, Shin SS, Han LL, Llanos F, Stowell M, Sloutsky A. Validation of a rapid method for detection of M. tuberculosis resistance to isoniazid and rifampin in Lima, Peru. Int J Tuberc Lung Dis 2005;9(7):760–4.

[11] Musa HR, Ambroggi M, Souto A, Angeby KA. Drug susceptibility testing of *Mycobacterium tuberculosis* by a nitrate reductase assay applied directly on microscopy-positive sputum samples. J Clin Microbiol 2005;43(7):3159–61.

[12] WHO. Rapid Implementation of the Xpert MTB/RIF diagnostic test. Technical and operational 'how-to' practical considerations. Geneva, Switzerland: World Health Organization; 2011. <https://www.who.int/tb/publications/tb-amplificationtechnology-implementation/en/> [accessed 11.07.19].

[13] WHO. Xpert MTB/RIF assay for the diagnosis of pulmonary and extrapulmonary TB in adults and children. Policy update. Geneva, Switzerland: World Health Organization; 2013. <https://www.who.int/tb/publications/xpert-mtb-rif-assay-diagnosis-policy-update/en/> [accessed 11.07.19].
[14] WHO. WHO Meeting report of a technical expert consultation, non-inferiority analysis of Xpert MTB/RIF Ultra compared to Xpert MTB/RIF. Geneva, Switzxerland: World Health Organization; 2017. <http://www.who.int/tb/publications/2017/XpertUltra/en/> [accessed 11.07.19].
[15] WHO. The use of molecular line probe assays for the detection of resistance to isoniazid and rifampicin. Policy update. World Health Organization; 2016. <https://www.who.int/tb/publications/molecular-test-resistance/en/> [accessed 11.07.19].
[16] WHO. The use of molecular line probe assays for the detection of resistance to second-line anti-tuberculosis drugs. Policy guidance. World Health Organization; 2016. <https://www.who.int/tb/publications/lpa-mdr-diagnostics/en/> [accessed 11.07.19].
[17] Su M, Satola SW, Read TD. Genome-based prediction of bacterial antibiotic resistance. J Clin Microbiol 2019;57:e01405−18.
[18] Meehan CJ, Goig GA, Kohl TA, Verboven L, Dippenaar A, Ezewudo M, et al. Whole genome sequencing of *Mycobacterium tuberculosis*: current standards and open issues. Nat Rev Microbiol 2019;. Available from: https://doi.org/10.1038/s41579-019-0214-5.
[19] WHO. Technical guide. The use of next-generation sequencing technologies for the detection of mutations associated with drug resistance in *Mycobacterium tuberculosis* complex. Geneva, Switzerland: World Health Organization; 2018. <https://www.who.int/tb/publications/2018/WHO_technical_guide_nextgen_sequencing/en/> [accessed 11.07.19].
[20] Cabibbe AM, Sotgiu G, Izco S, Migliori GB. Genotypic and phenotypic M. tuberculosis resistance: guiding clinicians to prescribe the correct regimens. Eur Respir J 2017;50:1702292.
[21] Consortium CR, the GP, Allix-Beguec C, Arandjelovic I, Bi L, Beckert P, et al. Prediction of susceptibility to first-line tuberculosis drugs by DNA sequencing. N Engl J Med 2018;379:1403−15.
[22] Madrazo-Moya CF, Cancino-Muñoz I, Cuevas-Cordoba B, Gonzalez-Covarrubias V, Barbosa-Amezcua M, Soberon X, et al. Whole genomic sequencing as a tool for diagnosis of drug and multidrug-resistance tuberculosis in an endemic region in Mexico. PLoS One 2019;14(6):e0213046.
[23] Quan TP, Bawa Z, Foster D, Walker T, del Ojo Elias C, Rathod P, et al. Evaluation of whole-genome sequencing for mycobacterial species identification and drug susceptibility testing in a clinical setting: a large-scale prospective assessment of performance against line probe assays and phenotyping. J Clin Microbiol 2018;56:e01480-17.
[24] Votintseva AA, Bradley P, Pankhurst L, del Ojo Elias C, Loose M, Nilgiriwala K, et al. Same-day diagnostic and surveillance data for tuberculosis via whole-genome sequencing of direct respiratory samples. J Clin Microbiol 2017;55:1285−98.
[25] Colman RE, Anderson J, Lemmer D, Lehmkuhl E, Georghiou SB, Heaton H, et al. Rapid drug susceptibility testing of drug-resistant *Mycobacterium tuberculosis* isolates directly from clinical samples by use of amplicon sequencing: a proof-of-concept study. J Clin Microbiol 2016;54:2058−67.
[26] Tagliani E, Hassan MO, Waberi Y, De Filippo MR, Falzon D, Dean A, et al. Culture and next-generation sequencing-based drug susceptibility testing unveil high levels of drug-resistant-TB in Djibouti: results from the first national survey. Sci Rep 2017;7(1):17672 pmid:29247181.

CHAPTER 6

Treatment guidelines for tuberculosis and tuberculous meningitis

Fiona V. Cresswell[1,2,3], Abdu K. Musubire[2,4] and Katarina M. Johansson Århem[5]
[1]Clinical Research Department, London School of Hygiene and Tropical Medicine, London, United Kingdom
[2]Infectious Diseases Institute, College of Health Sciences, Makerere University, Kampala, Uganda
[3]MRC-UVRI-LSHTM Uganda Research Unit, Entebbe, Uganda
[4]Department of Medicine, School of Medicine, College of Health Sciences, Makerere University, Kampala, Uganda
[5]Department of Infectious Diseases, Karolinska University Hospital, Stockholm, Sweden

Key points

- Tuberculous meningitis (TBM) is a highly fatal or disabling condition that requires careful monitoring and a high standard of supportive care.
- The optimal antimicrobial regimen for TBM is yet to be determined but currently the same regimen as for pulmonary tuberculosis is used with the continuation phase extended to a total treatment duration of 9–12 months.
- Research is underway to determine whether adjunctive drugs with better central nervous system penetration or higher doses of rifampicin can improve outcomes.
- Controlling immunopathology is also critical and currently corticosteroids are recommended; however, other agents, for example, aspirin, may be beneficial and are under investigation.
- Drug–drug interactions must be considered when rifampicin is used.

Drug-susceptible tuberculosis treatment

Why we use the current combination antituberculosis therapy

The first effective antibiotics against *Mycobacterium tuberculosis* (MTB) began use in the mid-1940s with the injectable drug streptomycin and oral *para*-aminosalicylic acid, PAS. In the beginning of the 1950s another more effective and less side effect prone oral agent was discovered, isoniazid [1]. Very soon it was realized that the use of a single agent quickly led to the development of MTB resistance; however, a combination of three drugs

DOI: https://doi.org/10.1016/B978-0-12-818825-5.00006-1

could cure the patient and this triple combination became standard treatment for more than a decade. The treatment in the beginning was long, up to 2 years for pulmonary tuberculosis (TB), with months spent in hospital because of the intravenous injections. Only in the late 1960s was an effective all-oral combination therapy developed, consisting of the newly discovered rifampicin and the rediscovered pyrazinamide together with isoniazid. This meant the course could be dramatically shortened to 6 months and ambulatory.

For many decades, no further advances were made for the cure of TB. The same combination therapy is still the standard for drug-susceptible TB worldwide, most often with the addition of ethambutol to cover for potential resistance. The course consists of an intensive phase followed by a continuation phase of different durations for different types of TB. Currently, there is a "one size fits all" approach of 2 months of rifampicin (R), isoniazid (H), pyrazinamide (Z), and ethambutol (E) followed by 4 months of R and H for drug-susceptible TB, except in TB meningitis and pericarditis where the recommended total duration is usually 9–12 months [2].

In most countries and settings, daily administration of fixed-dose combinations is used to improve adherence, reduce pill burden, and make dispensing easier. Thrice per week administration of drugs leads to increased risk of relapse and treatment failure and WHO 2017 guidelines no longer recommends this even in continuation phase. Four weight-bands are used to guide dosing (Table 6.1). In cases of side effects, toxicity, drug–drug interactions, or drug resistance, it may be necessary to stray from the fixed-dose combinations. In such circumstances, other first- or second-line anti-TB drugs can be used. These drugs are listed in Table 6.2 [3].

Improving tuberculosis therapy in the future

Since the 1970s, there was a long hiatus in research until TB started to increase again in the 1990s with the emergence of human immunodeficiency virus (HIV). Recently, there has been more research and investigation of repurposed drugs, higher doses of existing drugs, as well as new medicines in the pipeline. In 2012 bedaquiline, a diarylquinoline, became the first new anti-TB agent to be licensed in the United States since rifampicin in 1967 [4]. Delamanid and pretomanid, both nitroimidazoles, have been investigated for the treatment of multidrug-resistant TB (MDR-TB) [5,6]. In the future, customized therapy based on the extent

Table 6.1 Fixed-dose combination antituberculosis therapy.

Weight	Dose (daily) intensive phase	Dose (daily) continuation phase
	RHZ[a] **75/50/150 mg**	**RH** **75/50**
4–7 kg	1 tablet	1 tablet
8–11 kg	2 tablets	2 tablets
12–15 kg	3 tablets	3 tablets
16–24 kg	4 tablets	4 tablets
	RHZE **150/75/400/275 mg**	**RH** **150/75 mg**
25–37 kg	2 tablets	2 tablets
38–54 kg	3 tablets	3 tablets
55–70 kg	4 tablets	4 tablets
≥71 kg	5 tablets	5 tablets

E, Ethambutol; *H*, isoniazid; *R*, rifampicin; *Z*, pyrazinamide.
[a]Addition of ethambutol is recommended in the intensive phase for children with extensive disease or in settings with a high prevalence of HIV infection or isoniazid resistance.

and location of TB disease, HIV coinfection status, and other comorbidities would be an important step to improving outcomes. Baseline sputum smear grade of 3+ relative to < 2+, HIV seropositivity, and adherence of ≤ 90% are significant risk factors for unfavorable outcome. Hard-to-treat phenotypes of pulmonary TB, defined by high smear grades and cavitation, may require treatment durations of > 6 months for cure. Regimen duration can be selected in order to improve outcomes, provided that a stratified medicine approach is used instead of the "one-size-fits-all" treatment currently used worldwide, with an emphasis on proven methods to maximize adherence to treatment [7].

The current dosage of rifampicin was selected in the 1970s based on studies of pulmonary TB in the pre-HIV era. Pharmacokinetic studies have shown wide interindividual variations in plasma concentrations of both rifampicin and isoniazid. In addition, large individual differences in the activity of the *N*-acetyltransferase type 2 (NAT 2), the enzyme that metabolizes isoniazid, account for 10-fold variations in plasma concentration of the drug [8]. The reasons for interindividual variations of rifampicin concentrations in plasma are less well understood. Low body weight, male sex, and genetic polymorphisms have been associated with low plasma rifampicin concentrations [9]. Several recent and ongoing studies

Table 6.2 Antituberculosis drugs: dosing, side effects, and monitoring requirements.

Drug name	General description	Dose in children	Dose in adults	Renal failure	Penetration into CSF (%)	Side effects	Monitoring requirements and other comments
Rifamycins: Rifampicin (R, RIF) Rifabutin (Rfb)	Bactericidal, inhibits protein synthesis by blocking mRNA transcription and synthesis. Intracellular and extracellular action. Hepatically metabolized. Several studies imply that higher doses are safe and better outcome for both pulmonary TB and TB meningitis.	RIF: 15 (10–20) mg/kg, max 600 mg/day Rfb: Optimal dosing unknown. Estimated at 5–10 mg/kg	10 mg/kg Rfb: 5 mg/kg	No dose adjustment required For creatinine clearance <30 mL/min monitor concentrations	Rifampicin 20% Rifabutin 50%	*Common*: Orange-colored bodily secretions, hepatitis, GI-distress *Less common*: Cholestatic jaundice	*Monitoring*: Monitoring liver enzymes or for symptoms of DILI (nausea, jaundice, vomiting, and abdominal pain). *Comments*: Take on empty stomach or >2 h after food. Safe during pregnancy and breastfeeding. Rifampicin is a strong inducer of CYP450 and interacts with many other drugs Cross-resistance between rifamycins is around 20%.
Isoniazid (H, INH)	Fast bactericidal effect, especially early. Inhibits mycolic acid (cell wall) synthesis most effectively in dividing cells. Intra- and extracellular activity. Hepatically metabolized. There are large individual variations of concentration depending on if the individual's enzyme NAT2 has fast or slow activity. Increased dose may be of advantage in TBM but remains to be proven.	10 (10–15) mg/kg, max 300 mg/day	5 mg/kg High-dose H 16–20 mg/kg	No dose adjustment required	80%–90%	*Common*: Hepatitis, peripheral neuropathy (increased risk with malnutrition, pregnancy, alcoholism, diabetes, concurrent use of aminoglycosides, or ethionamide, HIV) *Less common*: Gynecomastia, rash, psychosis, and seizure	*Monitoring*: Consider liver enzyme monitoring, especially if age greater than 50 years. *Comments*: Give with pyridoxine (vitamin B6) to risk groups. Take on empty stomach or >2 h after food. Might interact with antiepileptics, phenytoin, and carbamazepine by inhibiting their metabolism in the liver. Safe during pregnancy and breastfeeding.
Pyrazinamide (Z, PZA)	Bactericidal. Mechanism unclear. Intracellular activity only. Effective in acidic milieu (e.g., cavitary disease). Hepatically metabolized, renally excreted.	35 (30–40) mg/kg	25 mg/kg	25 mg/kg/dose thrice weekly	90%–100%	*Common*: Gout, arthralgia, hepatotoxicity, and GI-distress *Less common*: Impaired diabetic control, rash	*Monitoring*: Liver enzyme monitoring, check uric acid if joint symptoms. *Comments*: Usually given once daily but can split dose initially to improve tolerance. Safe during pregnancy and breastfeeding.

Ethambutol (E, EMB)	Bacteriostatic at conventional dosing. Inhibits lipid and cell wall metabolism. Intra- and extracellular action. Renally excreted.	20 (15–25) mg/kg	15–25 mg/kg	15–25 mg/kg/dose thrice weekly	20%–30%	*Common*: Generally, well tolerated *Less common*: Optic neuritis, GI-distress, and arthritis/ arthralgia	*Monitoring*: Baseline and monthly visual acuity and red/green color vision test when dosed at greater than 15 mg/kg daily. Regularly question patient about visual symptoms. *Comments*: Safe during pregnancy and breastfeeding.
Aminoglycosides: Amikacin (Am) Kanamycin (Km) Streptomycin (S) Polypeptides: Capreomycin (Cm)	Bactericidal, inhibits protein synthesis through disruption of ribosomal function. Extracellular activity, less effective in acidic, intracellular environments. Polypeptides appear to inhibit translocation of the peptidyl-tRNA and the initiation of protein synthesis. Renally excreted. Long half-life, risk for toxic accumulation. Appears to have no role in improving outcome in TBM.	Am, Km, Cm: 15–30 mg/kg/day S: 20–40 mg/kg/day (max 1 g)	All: 15 mg/kg day Can be given intermittently after initial period	Usually 12–15 mg/kg 2–3 times per week. Therapeutic drug monitoring to avoid toxic accumulation	Am: 10%–20% Km: 10%–20% S: Low Cm unknown, likely to be low	*Common*: Pain at injection site, proteinuria, electrolyte wasting, cochlear ototoxicity resulting in hearing loss (related to cumulative and peak concentrations, increased risk with renal insufficiency, and can be irreversible) *Less common*: Nephrotoxicity (related to cumulative and peak concentrations, increased risk with renal insufficiency, often irreversible), peripheral neuropathy, rash, vestibular toxicity, and eosinophilia	*Monitoring*: Renal function tests and potassium. If potassium low, check magnesium and calcium. Baseline and monthly audiometry in high-risk patients (elderly, diabetic, HIV-positive, renal insufficiency). *Comments*: Monitor drug concentrations and reduce dose or increase dosing interval as needed. Ototoxicity may be potentiated by certain diuretics, especially loop diuretics. Avoid during pregnancy (potential ototoxicity in fetus) but ok during breastfeeding. Variable cross-resistance in this group have been reported.
Fluoroquinolones—moxifloxacin (Mfx), levofloxacin (Lfx)	Bactericidal, DNA-gyrase inhibitor. Well absorbed. Renally excreted	Mfx: Optimal dose unknown. Some experts use 10 mg/kg o.d., though lack of formulations makes such titration challenging. Aim for serum concentrations of 3–5 μL/mL 2 h post dose. Lfx: The optimal dose is not known but data suggest 13–20 mg/kg	Mfx: 400 mg daily Lfx: (500) 750–1000 mg daily	Mfx: No adjustment required Lfx: 750–1000 mg/dose thrice weekly for creatinine clearance <30 mL/min	70%–80%	*Common*: Generally considered well tolerated, but increasing concern about side effects, particularly musculoskeletal and nervous system *Less common*: Gastrointestinal, dizziness, photosensitivity, rash, tendonitis and tendon rupture, headache, insomnia, psychosis, seizures (more common in elderly), QT-prolongation	*Monitoring*: No laboratory monitoring required, check for clinical symptoms. *Comments*: Do not administer with antacids, sucralfate, iron, zinc, calcium, oral potassium, and magnesium replacements. Avoid in pregnancy and breastfeeding if possible, potential risk of arthropathy.

(*Continued*)

Table 6.2 (Continued)

Drug name	General description	Dose in children	Dose in adults	Renal failure	Penetration into CSF (%)	Side effects	Monitoring requirements and other comments
Cycloserine (Cs) and terizidone	Bacteriostatic, alanine analog, interferes with cell wall proteoglycan synthesis. Renally excreted. Terizidone (containing two molecules of cycloserine) can be used interchangeably in pulmonary TB, although there is limited evidence in TBM.	10–20 mg/kg/day (divided into 2 doses, max daily dose 1 g)	10–15 mg/kg/day divided or single dose depending on tolerability and tablet size	250 mg daily or 500 mg thrice/week or drug concentration monitoring	80%–90%	*Common*: Neurologic and psychiatric disturbances, including headaches, irritability, sleep disturbances, aggression, and tremors *Less common*: Psychosis, peripheral neuropathy, seizures (increased risk for with concurrent use of alcohol or other central acting drugs, isoniazid, ethionamide), hypersensitivity	*Monitoring*: Consider serum drug monitoring to establish optimal dosing. *Comments*: Give 50 mg pyridoxine, B6, for every 250 mg of cycloserine to lessen neurologic side effects. Consider dose-escalation over 2 weeks at treatment start. Not well studied in pregnancy, but no teratogenicity documented. Can be used when breastfeeding, give B6 supplementation to baby.
Tiamides: Ethionamide (Eto), protionamide (Pto)	Bactericidal or bacteriostatic depending on susceptibility and concentrations attained at infection site. Blocks mycolic acid synthesis. Hepatically metabolized, renally excreted. Ethionamide used successfully as a replacement for ethambutol to enhance TBM regimens in children. Prothionamide can be used interchangeably in pulmonary TB, limited evidence in TBM.	15–20 mg/kg/day (divided into 2–3 doses, max daily dose 1 g)	15–20 mg/kg/day divided or single dose	No adjustment required	80%–90%	*Common*: Gastrointestinal, dysgeusia (metallic taste) hypothyroidism (especially when taken with PAS) *Less common*: Arthralgia, dermatitis, gynecomastia, hepatitis, impotence, peripheral neuropathy, photosensitivity	*Monitoring*: Baseline liver enzymes. Consider monitoring TSH, especially if used with PAS. *Comments*: May split dose or give at bedtime to improve tolerability. Consider dose-escalation over 2 weeks at treatment start. Prothionamide has fewer GI side effects. Some cross-resistance with isoniazid. Avoid in pregnancy—reports of teratogenicity, little data on safety during breastfeeding. Usually recommended to use high-dose B6 supplementation.

Clofazimine (Cfz)	In vitro activity against *Mycobacterium tuberculosis* without much in vivo data. Generally reserved for cases with few other treatment options.	1 mg/kg/day (data limited)	100–200 mg daily	No adjustment required	No data, though has CNS side effects so likely to penetrate	*Common*: Pink or red discoloration of skin, conjunctiva, cornea, and body fluids. Gastrointestinal intolerance. Photosensitivity *Less common*: Retinopathy, dry skin, pruritus, rash, ichthyosis, xerosis, severe abdominal symptoms, including bleeding and bowel obstruction, QT-prolongation	*Monitoring*: Symptomatic monitoring. *Comments*: Not recommended during pregnancy or breastfeeding due to limited data.
Linezolid (Lzd)	In vitro bactericidal activity—increasing clinical experience. Inhibits protein synthesis. Good results in research for TBM.	10 mg/kg/dose every 12 h	600 mg once, or in some guidelines twice daily	No adjustment required. If hemodialysis, give dose after completion	70%	*Common*: Diarrhea and nausea *Less common*: Myelosuppression (with leukocytopenia, anemia, and thrombocytopenia), lactic acidosis. Optic and peripheral neuropathy (may be irreversible)	*Monitoring*: Check for optic neuritis and peripheral neuropathy. Complete blood count, weekly during the initial period then monthly. If symptoms of lactic acidosis, check lactic acid in blood. *Comments*: Pyridoxine is recommended. Do not use together with serotonergic drugs, such as MAOIs, SSRIs, and lithium as it might cause serotonin syndrome. Avoid or monitor closely if taken with tricyclic antidepressants. Not recommended during pregnancy and breastfeeding due to limited data.
Bedaquiline (Bdq)	Inhibits ATP-synthesis. Appears to have poor penetration into CSF, though data are limited, and binding to tubing make drug measurement following lumbar puncture challenging.	Not established	400 mg daily for 2 weeks then 200 mg thrice weekly for 22 weeks	No adjustment in mild-to-moderate renal failure	Low	*Common*: GI-distress, joint pain, and headache *Less common*: QT-prolongation, hyperuricemia, phospholipidosis (accumulation of phospholipids in body tissues), elevated aminotransferases, chest pain, and hemoptysis. Possible pancreatitis	*Monitoring*: Monitoring QT-interval with ECG, proposedly at baseline then weeks 2, 12, and 24 (more often if risk is elevated, e.g., when used together with other drugs that prolong QT—moxifloxacin, clofazimine, and antifungals). Discontinue if QT-interval exceeds 500 ms. Monitor liver enzymes. *Comments*: Not recommended during pregnancy and breastfeeding, limited data.

(*Continued*)

Table 6.2 (Continued)

Drug name	General description	Dose in children	Dose in adults	Renal failure	Penetration into CSF (%)	Side effects	Monitoring requirements and other comments
Nitroimidazoles: Delamanid (Dlm) Pretomanid (Pa)	Blocking synthesis of mycolic acids. Achieves high CNS concentrations in brain tissue in healthy rats, but brain or CSF distribution in infected animals or humans has not been assessed.	Delamanid: <5 years: no data 6–13 years: 50 mg b.d. >13 years: 100 mg twice daily	Delamanid: 100 mg twice daily	No adjustment in mild-to-moderate renal failure	Unknown	*Common*: Headache, dizziness, and nausea *Less common*: QT-prolongation	*Comments*: Passes over blood–placenta barrier and over to breast milk, but no studies on safety.
PAS	Bacteriostatic, disrupts folic acid metabolism. Hepatic acetylation, renally excreted.	200–300 mg/kg/day divided in 2–4 doses	8–12 g/day divided 2–3 times per day. Some experts use 6 g daily	No adjustment required	Unknown	*Common*: GI-distress, hypothyroidism (especially if combined with Eto) *Less common*: Hepatitis, electrolyte abnormalities, and coagulopathy	*Monitoring*: Consider monitoring TSH especially if taken together with Eto. *Comments*: Consider dose-escalation over 2 weeks at treatment start. PASER consists of granules that need to be taken with acidic food, such as yogurt or acidic juice. Drug interactions: decreased H acetylation, decreased R absorption in nongranular preparation, and decreased vitamin B12 uptake. Safety in pregnancy not studied, little data on breastfeeding.

CNS, Central nervous system; *CSF*, cerebrospinal fluid; *DILI*, drug-induced liver injury; *NAT2*, *N*-acetyltransferase 2; *PAS*, para-aminosalicylic acid; *TB*, tuberculosis; *TBM*, tuberculous meningitis; *GI*, gastrointestinal; *TSH*, thyroid-stimulating hormone; *MAOIs*, monoamine oxidase inhibitors; *SSRIs*, selective serotonin reuptake inhibitors.

Source: Adapted from USAID TB CARE II. The PIH guide to the medical management of multidrug-resistant tuberculosis 2nd edition. In: Partners in Health. Boston, MA: USAID TB CARE II; 2013. <https://www.pih.org/practitioner-resource/pih-guide-to-the-medical-management-of-multidrug-resistant-tuberculosis-2nd> [accessed 30.09.19], Curry International Tuberculosis Center and California Department of Public Health. Tuberculosis drug information guide. 2nd ed. 2012. <http://www.currytbcenter.ucsf.edu/products/tuberculosis-drug-information-guide-2nd-edition/main-page> [accessed 30.09.19], Cresswell FV, Te Brake L, Atherton R, Ruslami R, Dooley KE, Aarnoutse R, et al. Intensified antibiotic treatment of tuberculosis meningitis. Expert Rev Clin Pharmacol 2019;12(3):267–88, and WHO. Treatment of tuberculosis: guidelines. 4th ed. Geneva: World Health Organisation; 2010. <https://www.who.int/tb/publications/2010/9789241547833/en/> [accessed 30.09.19].

are looking at intensified rifampicin treatment. Higher doses seem to be safe and there is some evidence of better outcomes and faster clearance of TB [10–12]. Rifampicin doses could be individualized after achieving target drug concentration in plasma, at least in high-income settings. Several trials on shortening treatment to 4 months with the addition of fluoroquinolones have not been successful: despite earlier sputum conversion, there was a higher relapse rate [13,14]. A trial of high-dose rifampicin to shorten treatment is currently underway (NCT02581527).

Treatment of tuberculous meningitis

Treatment of tuberculous meningitis (TBM) is more challenging than pulmonary TB and this is mirrored in the wide differences in official guidelines. There is a lack of basic knowledge of both host and bacterial factors at play and of the pharmacokinetics and pharmacodynamics of the drugs when it comes to MTB infection of the central nervous system (CNS). The CNS is protected by the blood–brain barrier (BBB) and the blood-cerebrospinal fluid-barrier (BCSFB) that have several mechanisms that regulate the penetration of drugs into the CNS. There are both mechanisms that inhibit entry and mechanisms to expel drugs through different pumps (e.g., *P*-glycoprotein) and efflux systems, as shown in Fig. 6.1. Anti-TB drugs need to enter and stay at the site of disease for a sufficient time to kill the MTB bacilli. The distribution of bacilli in the CNS is often unknown and microscopic tuberculomas or tuberculous brain abscesses can exist alongside meningitis, so drugs must ideally penetrate brain tissue and cerebrospinal fluid (CSF). MTB can change its metabolic state over time which, in turn, can influence the efficacy of certain drugs depending on their mechanisms of action. It is currently unknown whether the different mechanisms of anti-MTB action (protein synthesis inhibition, disruption of cell wall synthesis, and bacterial enzyme inhibition) can influence host inflammatory responses and therefore vasculitis, strokes, and mortality [15].

Until recently there was a lack of pharmacometrics research of TB drugs in the CNS, largely because of the challenges of accessing the CNS. In the absence of clinical trial data to define the optimal treatment regimen for TBM, the current WHO-recommended regimen and most other guidelines have been adapted from recommendations for the treatment of pulmonary TB with an extended course of 9 or 12 months (Table 6.3) [2,16–20,22]. The prolongation of treatment is an empiric approach to

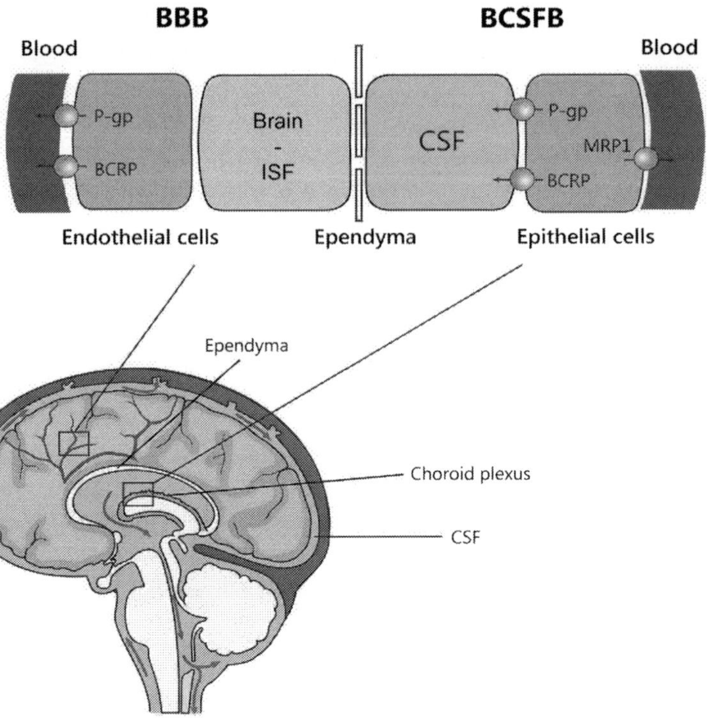

Figure 6.1 Schematic overview of the relation between the blood-brain barrier (BBB), made up by brain endothelial cells, and the blood-CSF barrier (BCSFB), made up by choroid plexus epithelial cells.

compensate for the recognized clinical severity of TBM, limited CNS penetration of some anti-TB drugs, undetected drug resistance, and patient nonadherence. Most deaths from TBM occur early in the treatment and relapse rates are not widely reported but are believed to be low [23]. A more reasonable strategy to intensify treatment and improve outcome would be to increase doses of poorly penetrating drugs and add drugs that better penetrate the BBB. General information on the range of anti-TB drugs and their ability to penetrate CSF are shown in Table 6.2.

Isoniazid and rifampicin are key components of treatment regimens for TBM. Isoniazid is a small and lipophilic drug that readily diffuses over the BBB and has potent early bactericidal activity. Rifampicin is large and protein bound and penetrates poorly over intact meninges but better in the earlier stages of inflammation. Even so the high mortality from rifampicin-resistant TBM has confirmed its central role in the treatment of CNS disease [24]. There is no conclusive evidence to demonstrate that

Table 6.3 Treatment guidelines for drug-susceptible tuberculous meningitis in adults.[a,b]

Guideline author	Recommendation	Website	References
WHO	2 months RHZS/4–10 months RH	https://www.who.int/tb/publications/2010/9789241547833/en/	[2,16]
IDSA/CDC	2 months RHZE/7–10 months RH	https://www.idsociety.org/practice-guideline/treatment-of-drug-susceptible-tb/	[17]
India	2 months RHZE/minimum 7 months RHE	https://apps.who.int/iris/handle/10665/278953	[18]
South Africa	2 months RHZE/7 months RH	https://health-e.org.za/2014/06/10/guidelines-national-tuberculosis-management-guidelines/	[19]
Uganda	2 months RHZE/10 months RH	http://library.health.go.ug/publications/tuberculosis/manual-management-tuberculosis-and-leprosy-uganda	[20]
China	No official guidelines, commonly used 3 months RHZE/9 months RHZ		[21]
United Kingdom	2 months RHZE/10 months RH	https://www.nice.org.uk/guidance/ng33/chapter/Recommendations#active-tb	[22]

[a] R—rifampicin, H—isoniazid, Z—pyrazinamide, E—ethambutol, S—streptomycin.
[b] All guidelines recommend adjunctive corticosteroids for 6–8 weeks.

pyrazinamide improves outcome of CNS TB but it is well absorbed orally and achieves high concentrations in the CSF. Therefore isoniazid, rifampicin, and pyrazinamide are considered mandatory in the intensive phase of TBM treatment. Some experts advocate the use of pyrazinamide throughout the whole course [18,21].

As an additional fourth drug, most authorities recommend either streptomycin or ethambutol, although neither penetrates the CSF well in the absence of inflammation and both can produce significant adverse reactions. Some centers, notably in South Africa, advocate ethionamide for children which penetrates healthy and inflamed meninges but it can cause severe nausea. Prothionamide may be better tolerated but is less well studied in TBM. The fluoroquinolones may represent an effective additional agent, although a large randomized controlled trial of adjunctive levofloxacin did not show a survival benefit over standard of care except in those with isoniazid mono-resistance [25]. Linezolid has shown promising results in a single observational TBM cohort in China [26]. Standard doses of rifampicin of 10 mg/kg seem to give lower than desirable concentrations in the CNS. There are several recent and ongoing studies exploring the safety and efficacy of higher doses of rifampicin in both pulmonary TB and TBM. Up to 35 mg/kg have been tried and well tolerated and higher doses may increase the early bacterial kill and improve survival [27–29].

Tuberculous meningitis in HIV-infected patients

Mortality from TBM is two- to three-fold higher in HIV coinfected individuals [30]. A number of studies show that HIV-positive patients achieve lower plasma concentrations of oral anti-TB drugs than HIV-negative patients, particularly in more advanced stages of HIV [31–33]. Possible explanations include HIV enteropathy with malabsorption, diarrhea, and wasting (risk of underdosing when using weight-based algorithms). HIV patients, due to decreased immune reactivity, appear to have less inflammation of the meninges and this may also result in poorer penetration of anti-TB drugs over the BBB. HIV-positive people are also at higher risk of side effects from anti-TB drugs as well as paradoxical reactions or immune reconstitution inflammatory syndrome (IRIS), and these reactions may be life threatening with an infection in the CNS. Following a study of immediate or delayed antiretroviral therapy (ART) in TBM which showed an increased in grade 4 adverse events (but not death) in the immediate ART arm [34], there is a general consensus that

ART-naive patients diagnosed with TBM should wait to start ART until week 8. However, while this approach may reduce the risk of TBM IRIS, it leaves a profoundly immunocompromised person at risk of other opportunistic infections for longer and there may be individual case scenarios where earlier ART is indicated. Important opportunistic infections that can both exist concomitantly and mimic the symptoms of CNS TB are cryptococcal meningitis and cerebral toxoplasmosis, so these should ideally be excluded in the diagnostic work-up.

For patients who already taking ART, drug–drug interactions and an overlapping toxicity with anti-TB agents can pose problems. For example, pyrazinamide, isoniazid, rifampicin, and efavirenz can all cause hepatotoxicity, and aminoglycosides and tenofovir can cause renal toxicity. While such drug toxicity is relatively easy to monitor and manage in higher income settings, in lower income settings with scarcity of laboratory tests and personalized drug regimens, it can pose a major management dilemma. As a result, treatment may need to be interrupted, which can lead to a worse outcome in TBM.

Antituberculosis drugs and antiretroviral therapy

Rifampicin is a strong inducer of the hepatic cytochrome P450 (CYP450) enzymes and coadministration leads to a decrease of concentrations of many other drugs, including many ART agents. The use of rifampicin with protease inhibitors (PIs) results in significant reduction of exposure resulting in HIV virological failure. For the treatment of pulmonary TB in HIV coinfected patients, rifabutin is recommended instead of rifampicin as it does not induce CYP450 to the same extent but there are little data on the use of rifabutin in TBM. Integrase inhibitors (INSTIs) such as raltegravir and dolutegravir require dose adjustment with increased pill burden and possible increase in toxicity. Of the non-nucleoside reverse transcriptase inhibitors (NNRTIs), nevirapine is contraindicated due to a marked reduction in exposure with rifampicin coadministration. Efavirenz levels are also reduced, though the reduction is not thought to be clinically significant. For more detailed drug–drug interaction information, the reader is referred to the Liverpool HIV Drug Interaction website (www.hiv-druginteractions.org) and summary in Table 6.4.

Tuberculous meningitis in pregnancy and breastfeeding

All first-line anti-TB drugs are safe in pregnancy and breastfeeding, with the exception of aminoglycosides, which could be ototoxic. Isoniazid,

Table 6.4 Summary of drug–drug interactions between commonly used antiretroviral and antituberculosis agents.

Class		NRTIs						NNRTIs				INSTIs			PIs					
		TDF	TAF	3TC	FTC	Abacavir	AZT	Efavirenz	Nevirapine	Rilpivirine	Etravirine	Dolutegravir	Raltegravir	Bictegravir	Ritonavir	Darunavir	Atazanavir	Lopinavir	Cobicistat	Maraviroc
First-line	Rifampicin	♦	■[b]	♦	♦	▲[u*]	■[c]	■[d]	●1	●1	●1	■[m]	■[m]	●3	●4	●4	●4,5	●6	●4	■[s]
	Rifabutin	♦	■[b]	♦	♦	♦	♦	■[e]	▲[w]	●2	■[k]	♦	♦	●3	■[o]	■[o]	■[o]	■[o]	■[o*]	■[t]
	Rifapentine	♦	■[b]	♦	♦	♦	♦	■[f*]	■[i]	●1*	●1*	■[n]	■[n]	●3*	■[p*]	■[p*]	■[p*]	■[p*]	●4*	■[p*]
	Isoniazid	♦	♦	♦	♦	♦	♦	♦	♦	♦	♦	♦	♦	♦	♦	♦	♦	♦	♦	♦
	Pyrazinamide	♦	♦	♦	♦	♦	♦	♦	♦	♦	♦	♦	♦	♦	♦	♦	♦	♦	♦	♦
	Ethambutol	♦	♦	♦	♦	♦	♦	♦	♦	♦	♦	♦	♦	♦	♦	♦	♦	♦	♦	♦
Second-line	Moxifloxacin	♦	♦	♦	♦	♦	♦	■[g*]	♦	▲[j]	■[g]	♦	♦	♦	♦	♦	■[q]	■[q]	♦	♦
	Ofloxacin/Levofloxacin	♦	♦	▲[v*]	♦	♦	♦			▲[j]	♦	♦	♦	♦	♦	♦	■[q]	■[q]	♦	♦
	Bedaquiline	♦	♦	♦	♦	♦	♦	■[h]	♦	■[j*]	■[l]	♦	♦	♦	■[r]	■[r]	■[r]	■[r]	■[r]	♦
	Pretomanid/Delamanid	♦	*	♦*	♦	♦*	♦*	♦	*	*	*				▲[x]	▲[x*]	▲[x*]	▲[x]		*
	Aminoglycosides	■[a*]	♦	▲[a*]	♦	♦	♦	♦	♦	♦	♦	♦	♦	♦	♦	♦	♦	♦	♦	♦

● Strong interactions

1. Rifamycins are a potent inducer of cytochrome P450 (CYP450) enzymes. The magnitude of rifapentine-mediated CYP3A4 induction is predicted to be lower than rifampicin but higher than rifabutin. Nevirapine, rilpivirine, and etravirine should not be used with rifampicin as coadministration may cause significant decreases in NNRTI concentrations and loss of therapeutic effect. No formal interaction studies with rifapentine yet.
2. Coadministration is contraindicated in the US Prescribing Information, but the European SPC recommends that the dose of rilpivirine should be increased to 50 mg o.d. during coadministration (and decreased to 25 mg o.d. when rifabutin is stopped).
3. Coadministration of rifampicin and rifabutin with a single dose of bictegravir (75 mg stat) decreased bictegravir C_{max} and AUC. Coadministration not recommended.
4. Coadministration contraindicated as it may significantly decrease concentrations of the PI, leading to loss of therapeutic effect and possible development of resistance.
5. Coadministration of twice daily atazanavir alone with rifampicin failed to provide adequate atazanavir exposure and a high frequency of liver reactions was seen.
6. Adequate exposure to lopinavir/ritonavir may be achieved when 400/400 mg twice daily is used but this is associated with a higher risk of liver and gastrointestinal toxicity. Therefore this coadministration should be avoided unless judged strictly necessary.

■ Potential interactions

a. Coadministration with drugs that reduce renal function or compete for active tubular secretion may increase concentrations of either drug. Avoid with concurrent or recent use of a nephrotoxic agent. If unavoidable, renal function should be monitored weekly.
b. Rifamycins are inducers that may result in lower exposure of TAF. Coadministration is not recommended, but if unavoidable, TAF 25 mg b.d. may provide comparable exposures to those observed with TAF 25 mg o.d. in the absence of rifampicin.
c. Rifampicin significantly decreased AZT AUC (47%) and C_{max} (43%). This may result in a partial loss or total loss of efficacy of AZT.
d. Studies (in African and Asian populations) indicate either that there is no clinically significant effect of rifampicin on efavirenz exposure so most guidelines recommend that efavirenz is used at 600 mg o.d. In the absence of efficacy data, patients maintained on efavirenz 400 mg o.d. should increase to efavirenz 600 mg o.d. while treated with rifampicin.
e. Coadministration of rifabutin (300 mg o.d.) and efavirenz (600 mg o.d.) decreased rifabutin C_{max} (32%), AUC (38%), and C_{min} (45%). Efavirenz C_{min} decreased by 12%, but there was no change in C_{max} or AUC. Increase daily doses of rifabutin by 50%; consider doubling rifabutin doses in regimens where rifabutin is given two to three times a week. The clinical effect of dose adjustment has not been adequately evaluated. Individual tolerability and virological response should be considered when making the dose adjustment.

f. Possible decrease in efavirenz and rifapentine concentrations. The magnitude of rifapentine-mediated CYP3A4 induction is predicted to be lower than rifampicin but higher than rifabutin. Perform therapeutic drug monitoring for efavirenz and adjust dosage if needed.

g. Moxifloxacin is predominantly glucuronidated by UGT1A1. Efavirenz and etravirine induces UGT1A1 and therefore could potentially decrease moxifloxacin levels. Monitor the clinical response.

h. Coadministration of bedaquiline (400 mg stat) and efavirenz (600 mg o.d.) to 33 HIV/TB-negative subjects decreased bedaquiline AUC by 18% and had no effect on C_{max}. Efavirenz PK similar to historical data from HIV-infected subjects. A reduction in bedaquiline exposure may result in loss of activity, coadministration is not recommended.

i. Nevirapine is unlikely to significantly alter rifapentine PK. Perform therapeutic drug monitoring for nevirapine and adjust dosage if needed.

j. Caution when coadministered with other drugs that prolong the QTc interval, an additive effect on QT-prolongation cannot be excluded. If coadministration is necessary, clinical monitoring, including frequent ECG assessment, is recommended.

k. See website www.hiv-druginteractions.org.

l. Etravirine may reduce bedaquiline exposure due to induction of CYP3A4, resulting in loss of activity.

m. Dose adjustment of dolutegravir to 50 mg b.d. and raltegravir to 400 mg b.d. is recommended (provided there is no underlying integrase class resistance in which case coadministration not recommended).

n. The magnitude of rifapentine-mediated CYP3A4 induction is predicted to be lower than with rifampicin but higher than with rifabutin. Use with caution until further data are available or consider using rifabutin.

o. Initial PK studies in healthy volunteers showed a significant increase in rifabutin with PI coadministration. Thus a reduction of rifabutin dosage to 150 mg three times a week was recommended to reduce the risk of rifabutin-related toxicity. However, more recent PK data derived from HIV/TB coinfected patients have shown that the coadministration of atazanavir/ritonavir (300/100 mg o.d.) and rifabutin (150 mg thrice weekly) resulted in rifabutin concentrations below the therapeutic levels suggesting that rifabutin dosage may be inadequate. Of interest, cases of relapses with acquired rifamycin-resistant MTB infection have been described in coinfected patients treated with rifabutin 150 mg thrice weekly and atazanavir/r or lopinavir/r. The US guidelines for HIV treatment now recommend the administration of rifabutin at a daily dosage of 150 mg with a boosted PI. Due to the limited safety data, patients should be closely monitored for rifabutin-related toxicities (i.e., uveitis or neutropenia).

p. Coadministration has not been studied and is not recommended as it may significantly decrease ritonavir concentrations, which may reduce the therapeutic effect. Consider using rifabutin.

q. Quinolones are metabolized by glucuronidation: inhibition of this pathway by atazanavir/lopinavir may increase moxifloxacin concentrations and prolong the QT-interval.

r. Bedaquiline is metabolized by CYP3A4 and moderate or strong CYP3A4 inhibitors may increase bedaquiline exposure, which could potentially increase the risk of adverse reactions. Bedaquiline prolongs the QTc interval.

s. Increase maraviroc to 600 mg twice daily.

t. In the *absence of a PI* and maraviroc should be administered at 300 mg twice daily with rifabutin. With rifabutin in the *presence of a PI*, the maraviroc dose should be decreased to 150 mg twice daily.

▲Potential weak interactions

u. Slight decrease abacavir levels due to induction of UGTs. No a priori dose adjustment required.

v. In vitro data indicate that levofloxacin and ofloxacin inhibits OCT2 and could potentially increase 3TC.

w. Possible increase in rifabutin levels. No adjustment required.

x. Partially metabolized by CYP3A4. ~20% Increased concentrations of delamanid with lopinavir/ritonavir.

3TC, Lamivudine; *ART*, antiretroviral; *AUC*, area under the time concentration curve; *AZT*, zidovudine; *b.d.*, twice daily; C_{max}, maximum (plasma) concentration; C_{min}, minimum (plasma) concentration; *ECG*, electrocardiogram; *FTC*, emtricitabine; *INSTI*, integrase inhibitors; *NNRTI*, non-nucleoside reverse transcriptase inhibitor; *NRTI*, nucleoside reverse transcriptase inhibitor; *o.d.*, once daily; *PI*, protease inhibitor; *PK*, pharmacokinetic; *SPC*, summary of product characteristics; *TAF*, tenofovir-AF; *TDF*, tenofovir-DF.

*Key: Coadministration has not been formally studied in human trials.

♦, No interaction expected; ▲ **a,** Potential interactions; ▲ **j,** Potential weak interactions.

Source: Information derived from www.hiv-druginteractions.org.

rifampicin, ethambutol, and pyrazinamide all cross the placenta—blood barrier, which, since the fetus also can develop congenital TB, is not a disadvantage. The risk for adverse effects such as drug-induced hepatitis and peripheral neuropathy, however, is increased in pregnant women. Pyridoxine (vitamin B6) supplementation should be given to all pregnant and breastfeeding women taking isoniazid. The safety of the second- and third-line drugs in pregnancy and breastfeeding is generally less studied and in many cases the risk of the mother must be weighed against the potential teratogenicity. For more detailed information on individual drugs see Table 6.2.

There is little known about changes of the disposition of anti-TB drugs in the pregnant body. One small pharmacokinetics study on rifampicin in third semester pregnant women with pulmonary TB and HIV in South Africa suggests that the concentrations seem to be similar to those who are not pregnant and dose adjustment is not necessary [35]. The recommended doses and duration of treatment are the same as for non-pregnant patients in all guidelines. In high-income settings, pregnant women with TB are usually followed in specialized antenatal care clinics for potential problematic pregnancies.

There have been some discussions on the safety of corticosteroids in pregnancy. Most studies have been done with low doses of corticosteroids for rheumatic diseases. There are some indications that intrauterine growth retardation as well as premature birth might be associated with corticosteroids, but no conclusive proof of abnormalities in children [36]. However, the benefit of using corticosteroids in pregnant women with TBM likely outweighs the potential side effects and should not be withheld.

Tuberculous meningitis in children

Because there are few clinical trials of antimicrobial treatment in pediatric TBM, the best treatment practices have been extrapolated largely from adult trials with doses adjusted for differences in drug disposition in children. The WHO-recommended pediatric dosing (Table 6.1) of first-line fixed-dose combination drugs was revised in 2010. Still, with these revised recommendations, a low proportion of children treated for pulmonary TB achieve therapeutic drug concentrations [37,38], which means that optimal treatment of TBM or CNS TB may require higher doses of the most life-saving drugs, most notably rifampicin. Two clinical trials are underway to inform the optimal treatment of pediatric TBM (NCT02958709, ISRCTN40829906). Based on adult pharmacokinetic (PK) targets for rifampicin associated with improvements in mortality and knowledge of the developmental pharmacology of this drug, rifampicin should probably be dosed at 30 mg/kg or higher in children. Second-line drugs have not been examined for efficacy, specifically in pediatric TBM. Linezolid was associated with improved clinical outcomes in adults in one study and could also have a beneficial role in pediatric TBM [39].

The duration of treatment has just as with adults remained controversial. According to the American Thoracic Society/Center for Disease Control and Prevention (CDC)/Infectious Diseases Society of America [17], a 9–12-month regimen is recommended. The British Infection Society [40] recommendations suggest a minimum of 12 months, whereas the WHO [41] recommends 12 months (2 months RHZE/10 months RH) of treatment to decrease the risk of disability and mortality. In some settings, treatment duration is shorter. In South Africa, for example, children are mostly treated for 6 months using high-dose isoniazid (20 mg/kg/day), high-dose rifampicin (20 mg/kg/day), standard-dose pyrazinamide (40 mg/kg/day), and ethionamide (20 mg/kg/day) (the latter in place of ethambutol) based on cohort data showing the efficacy and safety of this regimen [42]. The American Academy of Pediatrics is also recommending ethionamide instead of ethambutol for children, but with a conventional intensive/continuation phase [17]. The rational for using ethionamide as the fourth drug is that it has good CSF penetration and less adverse effects compared with those of streptomycin and ethambutol. Another advantage in many settings is that isoniazid mono-resistant TBM may be overcome when ethionamide and pyrazinamide are used continuously for the whole period.

Treatment of drug-resistant tuberculosis

Multidrug-resistant tuberculosis

In the early years of anti-TB therapy, drug resistance began to emerge as a significant challenge. The different degrees of resistance are characterized as follows: (1) mono-resistant TB, resistance to just one anti-TB drug; (2) MDR-TB, resistance to both isoniazid and rifampicin; (3) poly-resistant TB, resistance to more than one drug but not both isoniazid and rifampicin; and (4) extensively drug-resistant TB (XDR-TB), resistance to both isoniazid and rifampicin plus any fluoroquinolone and at least one injectable TB drug. Even though mono-resistant TB and other types of poly-resistant TB are more common than MDR-TB, the consequences of resistance to both rifampicin and isoniazid are grave. The second-line drugs are generally less effective, more toxic, and require longer treatment courses (see Table 6.2 for dosing) [43].

In most low- and middle-income settings, the fully automated nucleic acid amplification tests Xpert MTB/RIF and Xpert MTB/RIF Ultra (Cepheid, United States) are used for the rapid determination of rifampicin resistance (see Chapter 4: Laboratory methods for detecting tuberculosis and tuberculous meningitis, and Chapter 5: Identification of *Mycobacterium tuberculosis* drug resistance). Phenotypic culture-based drug susceptibility testing (DST) takes weeks and is not available in all settings (see Chapter 5: Identification of *Mycobacterium tuberculosis* drug resistance); therefore treatment has to be started using a standard first-line regimen. When the DST results are available (phenotypic or genotypic), treatment can be individualized. The traditional WHO MDR-TB treatment includes at least four effective core second-line anti-TB drugs, plus most often pyrazinamide, in the intensive phase, and three or more drugs in the continuation phase (see Table 6.5 for classification of anti-TB drugs and for standardized construction of an MDR-TB treatment regimen). If there are uncertainties about the likelihood of efficacy, the regimen may include more than five drugs in the intensive phase. The intensive phase is usually up to 8 months and the continuation phase is up to 12 months but can also be longer depending on the clinical response to treatment. Since 2016 the WHO has adopted the "Bangladesh model" all-oral regimen for some patient groups with pulmonary TB. This is a shorter treatment consisting of 4–6 months high-dose isoniazid, plus ethambutol and pyrazinamide, combined with four second-line drugs, followed by 5 months of ethambutol and pyrazinamide combined with two second-line drugs. However, this shorter regimen is currently not recommended for extrapulmonary TB. In 2019 the WHO published consolidated guidelines for the treatment of drug-resistant TB [44].

Multidrug-resistant tuberculous meningitis

MDR-TBM is almost always universally fatal unless the resistance is picked up early and treated in a specialist center. The best regimen for the treatment of MDR-TBM is unknown. Due to small patient numbers and delays in diagnosis of drug resistance, no randomized studies have been

Table 6.5 How to build a multidrug-resistant tuberculosis (MDR-TB) treatment regimen: WHO classification.

Step 1	Choose an injectable drug (Group 2)	Kanamycin (Km) or amikacin (Am) Capreomycin (Cm)	Base this on treatment history and DST. Streptomycin is generally not used because of high rates of resistance
Step 2	Choose a fluoroquinolone (Group 3)	Levofloxacin (Lfx) Moxifloxacin (Mfx)	
Step 3	Add at least two Group 4 drugs (oral bacteriostatic second-line agents)	PAS Cycloserine (Cs) or terizidone (Trd) Ethionamide (Eto) or protionamide (Pto)	Add Group 4 drugs until the regimen has at least four second-line drugs likely to be effective. Choice is based on treatment history and side effect profile
Step 4	Add Group 1 drugs (first-line oral anti-TB drugs)	Pyrazinamide (Z) Ethambutol (E)	Z is routinely added, except if the patient is intolerant or if resistance is highly likely based on history and DST. If the criterion of E being a "likely to be effective drug" is met it can be added, but not regarded a core drug
Step 5	Consider Group 5 drugs (agents with unclear role in treatment of drug-resistant TB)	Bedaquiline (Bdq) Delamanid (Dlm) Pretomanid (Pa) Linezolid (Lzd) Clofazimine (Cfz) Amoxicillin/clavulanic acid (Amx/Clv) Meropenem (Mpm) Imipenem/cilastatin (Ipm/Cln) Clarithromycin (Clr) Thioacetazone (Thz) High-dose isoniazid (high-dose H)	If there are less than four "likely to be effective" drugs from Groups 2 to 4, add at least two Group 5 drugs

DST, Drug susceptibility testing; PAS, para-aminosalicylic acid.

Source: Adapted from USAID TB CARE II. The PIH guide to the medical management of multidrug-resistant tuberculosis 2nd edition. In: Partners in Health. Boston, MA: USAID TB CARE II; 2013. <https://www.pih.org/practitioner-resource/pih-guide-to-the-medical-management-of-multidrug-resistant-tuberculosis-2nd> [accessed 30.09.19] and WHO. Treatment of tuberculosis: guidelines. 4th ed. Geneva: World Health Organisation; 2010. <https://www.who.int/tb/publications/2010/9789241547833/en/> [accessed 30.09.19].

performed and current practice is largely based on expert opinion. Treatment of individuals with MDR-TBM must always be discussed with an expert, including information on their DST results and treatment history. The British Infection Society recommends initiating therapy for MDR-TBM with at least a fluoroquinolone (either moxifloxacin or levofloxacin), pyrazinamide, ethionamide or prothionamide, and an injectable agent (amikacin or capreomycin), unless the susceptibility profile of the index case has shown resistance to any of these agents. Thereafter, treatment should be guided by national MDR experts [40].

Pyrazinamide shows a high penetration into CSF and is usually part of the MDR treatment regimen for pulmonary TB. It should be realized, however, that MDR-TB isolates are often resistant to pyrazinamide as well. Fluoroquinolones have good CNS penetration, moxifloxacin possibly better than levofloxacin. In TBM with isoniazid mono-resistance, clinical trial data have shown that use of adjunctive levofloxacin with a higher dose of rifampicin (15 mg/kg) can improve survival [12]. In line with MDR-TB treatment for pulmonary TB, a high dose of isoniazid (10−15 mg/kg daily) can be considered in cases of low-level resistance to isoniazid with mutations in the InhA promoter region outside of the katG gene.

Ethionamide, protionamide, and cycloserine all penetrate the CNS well, though these drugs carry high risk of toxicity and may only be tolerated for a short term. Ethionamide is recommended by the American Academy of Pediatrics for drug-susceptible TBM [17]. Linezolid is increasingly used in treatment of other multi-resistant bacterial CNS infections thanks to its excellent penetration over the BBB. A single observational study from China showed promising results when linezolid was used as a compliment to standard treatment in TBM [26]. However, there are some concerning side effects especially when used for longer periods and the optimal dose and treatment duration remains to be determined.

The carbapenems, meropenem, and imipenem have good penetration into the CNS but the extent of their anti-TB effect is currently not clear. The nitroimidazoles (e.g., delamanid or pretomanid) might play a role in the treatment of CNS TB. Animal studies suggest high concentration in the brain tissue, even though CSF concentrations appear to be low [10]. Aminoglycosides penetrate inflamed meninges to some extent and are an option in the intensive phase of treatment for drug-susceptible TBM so they could play a role in MDR-TBM.

Corticosteroids and host-directed therapies for tuberculous meningitis

TBM can cause a wide range of neurological disabilities in survivors. The intracerebral and spinal pathology in TBM is mediated by a dysregulated inflammatory response that contributes to meningitis, tuberculoma formation, arteritis, obstruction of CSF flow, and vascular complications, including stroke. To dampen this immunopathology, adjunctive corticosteroid treatment is generally recommended but has also been disputed. A 2016 Cochrane systematic review and meta-analysis on the use of adjunctive corticosteroids in TBM included seven randomized controlled trials involving 1140 participants (with 411 deaths). It concluded that corticosteroids improved outcome in HIV-negative children and adults with TBM, but as all but one of the studies excluded HIV-infected individuals, the benefit in this group still remains uncertain [45]. In HIV-infected patients, especially those with advanced disease, further immune suppression with corticosteroids may increase their vulnerability to other opportunistic infections. Also, with a less marked immune response because of the HIV infection itself, it is not certain that this group will benefit from corticosteroids to the same extent as HIV-negative individuals. The benefit of corticosteroids is likely to vary by stage of HIV infection and use of ART. A trial of corticosteroids or placebo in TBM in people living with HIV is currently underway in Vietnam [46].

Dexamethasone does not seem to affect the incidence or resolution of hemiparesis, paraparesis, or quadriparesis, which are the most common causes of severe disability due to TBM. It may therefore be hypothesized that corticosteroids exert an effect by reducing basal meningeal inflammation and brainstem encephalopathy but do not modify infarct-causing periarteritis. Other host-directed anti-inflammatory therapies may be needed to reduce vasculitis and infarcts. There are no data from controlled trials comparing different corticosteroid regimens; therefore the choice of regimen should be based on local feasibility and existing protocols. A commonly used regimen for adult patients with no coma or focal signs is dexamethasone 0.3 mg/kg/day (max 24 mg) tapered over 6 weeks and for patients with coma or focal signs, dexamethasone 0.4 mg/kg/day (max 24 mg), tapered over 8 weeks. For children a possible regimen could be prednisolone 2–4 mg/kg/day (or equivalent dose dexamethasone: 0.3–0.6 mg/kg/day) for 4 weeks, followed by a reducing course over 2–4 weeks [40,47].

Corticosteroids are imperfect at controlling the inflammatory milieu that results in cerebral infarction and subsequent disability. Other agents may be more effective, though this is a novel area of research and the best agent and in which population remains to be defined. A recent phase II trial of adjunctive aspirin in HIV-negative people with TBM in Vietnam showed some promising results in participants with microbiologically confirmed TBM in whom 1000 mg of aspirin daily reduced new infarcts and death by day 60 to 14%, compared to 34% in the placebo group [48]. CSF analysis showed aspirin dose-dependent inhibition of thromboxane A_2. Aspirin appeared to be well tolerated but larger studies and trials in HIV-positive patients are needed before aspirin can be recommended in TBM. A randomized clinical trial of thalidomide 28 mg/kg/day, a potent inhibitor of tumor necrosis factor-alpha (TNF-α), was conducted in children with TBM in South Africa but was halted early due to safety concerns, possibly related to the high dose used [49].

Complications of tuberculous meningitis

Mortality in TBM remains high, between 20% and 50%, despite treatment [50]. The high mortality and morbidity might result from the disease itself, its complications, or its treatment [51]. Delayed diagnosis is still an important factor because there is still no single diagnostic method that is both sufficiently rapid and sensitive to diagnose all cases of TBM. The only way to reduce the mortality and morbidity is by earlier diagnosis, reduction of pathogen or host-mediated immunopathology, and timely recognition of complications with institution of the appropriate interventions. Aggressive management of complications is thought to reduce secondary brain injury and improve outcomes. The most common complications include elevated intracranial pressure, hydrocephalus, vasculitis, seizures, and hyponatremia [52,53].

Fever

The prevalence of fever in TBM is 60%–95% [52]. Fever exacerbates neurological injury in the presence of a cerebral insult and also raises intracranial hypertension (ICP) [51]. The presence of fever is associated with increased 1-year mortality in HIV-negative patients with TBM [51]. Lowering the temperature lowers cerebral metabolic rate and oxygen demand but therapeutic hypothermia has not showed beneficial results [51]. Standard fever management consists of antipyretic drug therapy and external physical cooling [52]. Newer methods of surface-cooling and intravascular-cooling devices are more effective in decreasing fever than standard fever-management protocols. Achieving normothermia might be justified in TBM. However, one must exercise caution in patients with associated sepsis because hypothermia has been associated with reduced ability to clear infections, a factor that is disputed in animal studies [54].

Hyponatremia

The prevalence of hyponatremia ranges from 35% to 71% in patients with TBM [55]. Hyponatremia is an independent predictor of death or severe disability [56]. Brain swelling due to hyponatremia in children is less well tolerated than in adults, hence considerable morbidity from hyponatremia in the pediatric population. The severity of hyponatremia can be classified as severe ($Na^+ < 120$ mEq/L), moderate ($Na^+ = 120-129$ mEq/L), or mild ($Na^+ = 130-134$ mEq/L) [55]. It is important to asses for the volume status of all TBM patients at presentation, since most will present with hypovolemia and this may also be important in analyzing the cause of hyponatremia. Clinical presentation may include headache, confusion, and seizures [51]. The clinical parameters suggesting volume contraction such as dehydration, high urinary output, weight loss, and low central venous pressure (CVP) can be helpful.

The possible causes are cerebral salt-wasting syndrome (CSWS), syndrome of inappropriate secretion of antidiuretic hormone (SIADH), and adrenal insufficiency, though recently the most likely cause has been suggested to be CSWS [55]. CSWS is characterized by natriuresis resulting in hyponatremia and hypovolemia in response to brain injury. SIADH occurs due to excessive release of antidiuretic hormone from the posterior pituitary gland resulting in inappropriate, continued secretion, or action of the antidiuretic hormone arginine vasopressin despite normal or increased plasma volume leading to euvolemic or hypervolemic hyponatremia [52,57]. The difference between CSW and SIADH is blurred because both these conditions have low serum sodium and serum osmolality with high urinary osmolality and urinary sodium. A number of clinical conditions such as poor oral intake, extrarenal loss (vomiting, diarrhea), endocrine abnormalities, and heart failure have to be considered before diagnosing CSW or SIADH [56].

In one study by Misra et al. [55], CSWS was considered in the presence of at least two out of four of the following features in a patient with hyponatremia: (1) clinical findings of hypovolemia such as hypotension, dry mucous membranes, tachycardia, or postural hypotension; (2) laboratory evidence of dehydration such as elevated hematocrit, hemoglobin, serum albumin, or blood urea nitrogen; (3) negative fluid balance as determined by intake–output chart and/or weight loss; and (4) CVP below 6 cm of water. SIADH was diagnosed in the presence of at least two out of four following features in a patient with hyponatremia: (1) no signs of hypovolemia such as hypotension, dry mucous membrane, tachycardia, or postural hypotension; (2) no laboratory evidence of dehydration such as elevated hematocrit, hemoglobin, serum albumin, or blood urea nitrogen; (3) normal or positive fluid balance with the absence of weight loss; and (4) CVP above 6 cm of water [55].

Reduced serum osmolality, inappropriately high urine osmolality, elevated urinary sodium, and hypovolemia are consistent with CSWS, and 0.9% normal saline administration will be administered to correct circulating volume and serum sodium concentration [51]. Oral salt supplementation may be used in addition to the normal saline [55]. Hypertonic saline may also be used in symptomatic patients in severe cases provided appropriate monitoring is available [57]. Fludrocortisone has been used in salt-wasting syndrome especially in refractory cases and showed faster correction of the hyponatremia than normal saline alone without significant impact on hospital mortality, disability at both 3 and 6 months. However, this is limited by adverse events such as hypokalemia, hypertension, and pulmonary edema [58].

SIADH is managed by fluid restriction but this might result in worsening hypovolemia and harm in the context of an active infection [57]. Salt tablets and furosemide may be tried if the condition persists. In symptomatic severe cases, hypertonic saline 3% sodium chloride may be used. Vasopressin receptor antagonists, for example, tolvaptan and conivaptan, may also have a role in persistent hyponatremia [59,60]. The management of adrenal insufficiency is by administration of mineralocorticoid and glucocorticoid replacement [59]. It is important to remember the basic principles of correction of hyponatremia, including frequent monitoring of sodium levels, so as not to exceed a rate of correction of more than 10 mEq/L/24 h or 1 mEq/L every 2 hours to avoid central pontine myelinolysis, a potentially devastating demyelination of the brainstem and/or cerebral white matter, especially in cases of long-standing hyponatremia. In addition, avoid over correction that can lead to hypernatremia ($Na+ > 145$ mEq/L) which can cause muscle twitching, lethargy, confusion, and coma.

Seizures

Seizures in TBM have been reported in 17%–93% of patients and are categorized as per International League Against Epilepsy into early (within 1 month of illness) and late (after 1 month of illness) [61]. Seizures occur in 30% as early onset and 70% as late onset [61]. The seizures can be focal, focal to bilateral, generalized tonic–clonic, and status epilepticus. Seizure risk can be increased by drugs such as isoniazid or fluoroquinolone co-administration. Early seizures are associated with meningeal irritation and cerebral edema, while late seizures with tuberculoma, infarction, hydrocephalus, and hyponatremia [61].

Acute abortive treatment with benzodiazepines, for example, diazepam and lorazepam, is indicated for prolonged or repetitive seizures. In cases of status epilepticus, this is usually followed by a loading dose of phenytoin/fosphenytoin and subsequent maintenance therapy. Alternative medications for status epilepticus include phenobarbital, sodium valproate, levetiracetam, lacosamide, and midazolam [62]. Patients with acute symptomatic seizures in the setting of infections have a lower rate of unprovoked seizure recurrence over a 10-year time frame than that of patients with unprovoked remote symptomatic seizures [63]. Patients at risk of seizure recurrence include those with persistent epileptiform activity on EEG, structural changes on imaging, late-onset seizure in the course of disease, and status epilepticus during the acute phase. Antiepileptic drugs (AED) may be continued for a period of 3–6 months as there is a high risk of recurrence [64]. The incidence of seizure recurrence in TBM survivors is about 10% and these patients require long-term AED [65].

Isoniazid, rifampicin, pyrazinamide, and valproic acid are all potentially hepatotoxic drugs. If valproic acid is given, liver function should be monitored at regular intervals [52]. Isoniazid may increase the toxicity of carbamazepine and valproate, while rifampicin may reduce plasma levels of phenytoin [16]. Levetiracetam is a good option if available/affordable due to its low adverse side effect profile and lack of significant drug interactions. One should always consider nonconvulsive seizures and nonconvulsive status epilepticus in patients in a comatose state.

Raised intracranial pressure and hydrocephalus

Raised intracranial pressure (ICP) is associated with poor outcomes. In children with TBM, the relative risk for poor prognosis is two-fold.

Possible mechanisms include cytotoxic and vasogenic edema, vasculitis within the cerebral vessels and infarction, hydrocephalus, and space-occupying effects of associated tuberculoma(s) [51]. The presence of parenchymal pathology leads to failure of cerebral vascular autoregulation and metabolic abnormalities such as hyponatremia, hyperthermia, and hypercapnia can cause further dysregulation. Other players in the pathogenesis of raised ICP in TBM include fever, seizures, and impaired ventilation [51].

Hydrocephalus, the most common cause of raised ICP, is caused by the inflammatory infiltrate within the subarachnoid space or the ventricular pathways leading to disruption of CSF flow. It can be categorized as communicating hydrocephalus caused by abnormal CSF flow through the basal cisterns or the absorptive arachnoid granulations or non-communicating hydrocephalus due to CSF obstruction at the level of the ventricular aqueducts and/or fourth ventricle [51]. The presentation of hydrocephalus includes headache, vomiting, visual loss, and focal neurological deficits and clinical signs include papilledema and reduced level of consciousness. The investigation of choice is neuroimaging (ideally MRI) that can provide information about the severity and possible pathological causes of hydrocephalus. Differentiating non-communicating from communicating hydrocephalus is not always straightforward and some specialist neurosurgical centers employ air encephalography to make the diagnosis [51]. Non-communicating hydrocephalus is less common than communicating hydrocephalus but often requires surgical intervention (see Chapter 7: Neurosurgical management of tuberculous meningitis).

Monitoring intracranial pressure

The UK Medical Research Council TBM grading system combines the Glasgow Coma Scale and focal neurological signs to categorize disease severity, with grade 1 being mild disease and grade 3 being severe disease. A higher TBM grade at presentation predicts mortality regardless of HIV status. EEG monitoring is recommended in critically ill individuals with meningoencephalitis who are comatose to rule out nonconvulsive status epilepticus. Transcranial Doppler ultrasound can be used to determine cerebral blood flow in the basal arteries with intracranial pressure. Optic nerve sheath diameter ultrasound is a quick, easy, and safe method for measuring ICP, though this is yet to be validated in TBM. Lumbar puncture with CSF opening pressure measurement using a manometer can be used to quantify lumbar CSF pressure, which in most cases reflects cerebral CSF pressure unless there is non-communicating hydrocephalus or spinal arachnoiditis [51]. The gold standard for intracranial pressure monitoring can be done invasively at the intraventricular, intradural, and epidural sites but this is done with an increased risk of infection and hemorrhage and hence contraindicated in patients with coagulopathy and sites with local infection [51]. Biomarkers of brain injury are still under investigation as potential markers of severity of injury or for monitoring disease progress [51].

Management of raised intracranial pressure

Management of raised ICP has three components; supportive management, medical therapy, and surgical intervention [51]. Supportive care aims to optimize patient position (elevation of head of bed) and monitor/control parameters such as temperature, blood pressure, seizures, and hemoglobin levels [51]. Medical therapy involves correction of abnormalities in gas exchange and tissue oxygenation through mechanical ventilation (if necessary), meticulous fluid and electrolyte management, and use of osmotic diuretics in patients not requiring neurosurgical intervention. The commonly used osmotic agents are mannitol and hypertonic saline. The side effects of hypertonic saline include hypernatremia and osmotic blood–brain barrier opening with harmful extravasation of hypertonic saline solution in to the brain tissue. Mannitol carries a risk of dehydration and renal tubular damage. Hydrocephalus can be treated with the diuretics acetazolamide and furosemide, serial lumbar punctures, external ventricular drainage, ventriculoperitoneal shunting, or endoscopic third ventriculostomy (see Chapter 7: Neurosurgical management of tuberculous meningitis) [66]. Patients on medical treatment should be closely monitored for worsening or lack of improvement that would warrant surgical management. Early shunt surgery (2 days after diagnosis) has been associated with better outcomes compared to delayed surgery (3 weeks after diagnosis) [67].

Stroke

The incidence of stroke is about 13%–57% in TBM patients and is the main cause of long-term neurological disability [51]. The mortality is about three times higher in TBM patients with stroke compared to those without. Stroke is associated with the stage/severity of TBM, basal meningeal enhancement, hydrocephalus, exudate, and hypertension (see Chapter 3: Clinical presentations and features of tuberculous meningitis) [68]. Blood vessel pathology is mainly secondary to inflammation and necrosis secondary to basal exudates [51,69]. This vasculitis secondary to the meningeal inflammation can be classified into three patterns: infiltrative, proliferative, and necrotizing vascular lesions. Other vascular pathologies associated with TBM include arteritis, arterial spasm, intraluminal thrombus, and external compression of the proximal vessels with basal cistern exudates and compromise of cerebral perfusion and oxygen delivery to the brain [52]. Impaired cerebral perfusion leads to ischemia, cerebral infarction, and raised ICP. The other mechanisms involved may be the prothrombotic state and dehydration (hypovolemia).

The most common vulnerable brain region is the basal ganglia that has been described as the "tubercular zone" [70,71]. It is comprised of the head of the caudate nucleus, putamen, globus pallidus, anteromedial thalamus, and anterior limb and genu of internal capsule. Ischemia of the basal ganglia in TBM is attributed to the involvement of lenticulostriate, thalamotuberal, and thalamostriate arteries that are embedded in exudates and likely to be stretched by a coexistent hydrocephalus. Cortical stroke can also occur due to the involvement of proximal portion of the middle, anterior, and posterior cerebral arteries as well as the supraclinoid portion of the internal carotid and basilar arteries that are documented in MRI, angiography, and autopsy studies.

Hemiplegia is the most common clinical consequence of cerebral infarction due to TBM [52]. Other presentations will include seizures and altered sensorium. The consequences of irreversible neurological disability may present with feeding difficulties, pneumonia, pressure sores, and deep venous thrombosis. The lack of physiotherapy and rehabilitation services in many TB endemic settings increases the morbidity and mortality due to stroke. Patients are managed with standard anti-TB therapy, symptomatic treatments, and supportive care.

Tuberculomas

Tuberculomas can occur together with or independently of TBM. Clinical presentation depends on the site and includes seizures, focal neurological signs, and/or symptoms of raised ICP due to hydrocephalus or mass effect [72]. Tuberculomas commonly present as a feature of paradoxical worsening in patients treated for TB or in HIV-infected patients after initiation of ART (see Chapter 3: Clinical presentations and features of tuberculous meningitis). The mainstay of treatment remains anti-TB therapy, the duration of which is debated. There is lack of evidence in this area but consensus suggests four anti-TB drugs for 18 months or until the tuberculoma resolves on neuroimaging [73,74]. Surgical excision is indicated if (1) the tuberculoma is causing obstructive hydrocephalus and significant ICP or compartmental shifts and (2) the tuberculoma is not resolving or worsening on medical treatment. A number of case series suggest that thalidomide, anti-tumor necrosis factor biological agents (such as infliximab), and interferon-γ treatment may be beneficial [57].

Paradoxical reactions

Paradoxical reactions are when worsening signs and symptoms of TB occur despite effective anti-TB chemotherapy. These reactions are commonly considered an exuberant inflammatory response to dead or dying bacilli, though their pathogenesis is poorly understood. These reactions usually occur after 2–4 months of anti-TB chemotherapy. Paradoxical reactions contribute to critical illness of patients with TBM, with clinical features including headache, altered vision, and seizures. Neuroimaging findings include enhancing basal exudates, new or worsening tuberculomas, and optochiasmatic or spinal arachnoiditis. Paradoxical changes might cause mass effect and obstruct CSF with associated raised ICP. Clinical spinal disease is common and might be overlooked and is often revealed after commencing anti-TB chemotherapy [51]. IRIS occurs in the setting of HIV after introduction of ART [75,76]. Neurological manifestations of IRIS include meningitis, brain tuberculoma, brain abscess, radiculomyelitis, and spinal epidural abscess (see Chapter 3: Clinical presentations and features of tuberculous meningitis). Case reports and small case series suggest that adjunctive thalidomide, infliximab, and interferon-γ might be effective in situations where corticosteroids fail to improve clinical symptoms of IRIS [77,78].

Cerebral venous sinus thrombosis

Cerebral venous sinus thrombosis (CVST) is rarely reported and hence requires a high index of suspicion and timely intervention [79]. The prevalence in a magnetic resonance venography study among patients with TBM was found to be 11.2% [80]. The three main risk factors for CVST are stasis, endothelial inflammation, and a hypercoagulable state which are all present in TBM [80]. The hypercoagulable state may be due to the inflammatory response, platelet aggregation, and release of procoagulant factors. The other possible mechanism is direct compression of the sinus by granulation tissue or abscess [79]. Worsening of preexisting headache, impairment of consciousness, and seizures should raise suspicion of CVST in any patient with CNS infection [81]. The investigation of choice is a magnetic resonance venogram, although CT venography is often diagnostic and more widely available. The presence of filling defects in the sagittal, transverse, or sigmoid sinuses and non-visualization of the deep cerebral venous system are strongly suggestive of thrombosis in the presence of other supporting clinical or radiological information [80]. The recommended treatment is initial anticoagulation with heparin followed by oral anticoagulation for 3–6 months with caution due to an increased risk of bleeding [70,81]. Supportive treatments include seizure control and measures to decrease ICP and at times, decompressive craniectomy [79].

Drug-induced liver injury

Anti-TB drug-induced liver injury (DILI) is a significant concern, although the incidence varies widely depending on the patient population, drug combinations used, and comorbidities [82,83]. Almost all episodes occur in the first 3 months of anti-TB therapy (especially when R, H, and Z are given in combination), a critical time in TBM treatment. Currently, many physicians will withhold all TB treatment if hepatic transaminases rise above three to five times the upper limit of normal. This is probably a safe practice in pulmonary TB treatment, but in TBM first-line drug treatment interruptions are associated with an increased risk of death. The prognostic value of monitoring hepatic transaminases after the start of anti-TB treatment and the suggested thresholds for stopping anti-TB drugs are not evidence-based. However, outside of a clinical trial or well-resourced hospital setting, it can be challenging to monitor liver function tests regularly and caution must be taken not to cause long-term hepatic damage by being slow to take action in cases of serious DILI. If first-line anti-TB drugs need to be interrupted, patients can still receive some anti-mycobacterial therapy with a holding regimen of an aminoglycoside, a fluoroquinolone, and ethambutol pending reinitiation of first-line drugs [82,83].

References

[1] Chakraborty S, Rhee KY. Tuberculosis drug development: history and evolution of the mechanism-based paradigm. Cold Spring Harb Perspect Med 2015;5(8):a021147.

[2] WHO. Treatment of tuberculosis: guidelines. 4th ed. Geneva: World Health Organisation; 2010, Available from: https://www.who.int/tb/publications/2010/9789241547833/en/ [accessed 30.09.19].

[3] Curry International Tuberculosis Center and California Department of Public Health. Tuberculosis drug information guide. 2nd ed. <http://www.currytbcenter.ucsf.edu/products/tuberculosis-drug-information-guide-2nd-edition/main-page>; 2012 [accessed 30.09.19].

[4] Deoghare S. Bedaquiline: a new drug approved for treatment of multidrug-resistant tuberculosis. Indian J Pharmacol 2013;45(5):536–7.

[5] Liu Y, Matsumoto M, Ishida H, Ohguro K, Yoshitake M, Gupta R, et al. Delamanid: from discovery to its use for pulmonary multidrug-resistant tuberculosis (MDR-TB). Tuberculosis 2018;111:20–30.

[6] Barthod L, Lopez JG, Curti C, Bornet C, Roche M, Montana M, et al. News on therapeutic management of MDR-tuberculosis: a literature review. J Chemother 2018;30(1):1–15.

[7] Imperial MZ, Nahid P, Phillips PPJ, Davies GR, Fielding K, Hanna D, et al. A patient-level pooled analysis of treatment-shortening regimens for drug-susceptible pulmonary tuberculosis. Nat Med 2018;24(11):1708–15.

[8] Zabost A, Brzezinska S, Kozinska M, Blachnio M, Jagodzinski J, Zwolska Z, et al. Correlation of *N*-acetyltransferase 2 genotype with isoniazid acetylation in Polish tuberculosis patients. BioMed Res Int 2013;2013:853602.

[9] Verbeeck RK, Gunther G, Kibuule D, Hunter C, Rennie TW. Optimizing treatment outcome of first-line anti-tuberculosis drugs: the role of therapeutic drug monitoring. Eur J Clin Pharmacol 2016;72(8):905–16.

[10] Cresswell FV, Te Brake L, Atherton R, Ruslami R, Dooley KE, Aarnoutse R, et al. Intensified antibiotic treatment of tuberculosis meningitis. Expert Rev Clin Pharmacol 2019;12(3):267–88.

[11] Heemskerk AD, Bang ND, Mai NT, Chau TT, Phu NH, Loc PP, et al. Intensified antituberculosis therapy in adults with tuberculous meningitis. N Engl J Med 2016;374(2):124–34.

[12] Heemskerk AD, Nguyen MTH, Dang HTM, Vinh Nguyen CV, Nguyen LH, Do TDA, et al. Clinical outcomes of patients with drug-resistant tuberculous meningitis treated with an intensified antituberculosis regimen. Clin Infect Dis 2017;65(1):20–8.

[13] Jindani A, Harrison TS, Nunn AJ, Phillips PP, Churchyard GJ, Charalambous S, et al. High-dose rifapentine with moxifloxacin for pulmonary tuberculosis. N Engl J Med 2014;371(17):1599–608.

[14] Gillespie SH, Crook AM, McHugh TD, Mendel CM, Meredith SK, Murray SR, et al. Four-month moxifloxacin-based regimens for drug-sensitive tuberculosis. N Engl J Med 2014;371(17):1577–87.

[15] Palucci I, Delogu G. Host directed therapies for tuberculosis: futures strategies for an ancient disease. Chemother 2018;63(3):172–80.

[16] WHO. Guidelines for treatment of drug-susceptible tuberculosis and patient care (2017 update), <https://www.who.int/tb/publications/2017/dstb_guidance_2017/en/>; 2017 [accessed 30.09.19].

[17] Nahid P, Dorman SE, Alipanah N, Barry PM, Brozek JL, Cattamanchi A, et al. Executive summary: official American Thoracic Society/Centers for Disease Control and prevention/infectious diseases society of America Clinical Practice Guidelines: treatment of drug-susceptible tuberculosis. Clin Infect Dis 2016;63(7):853–67.

[18] WHO. Available from: https://apps.who.int/iris/handle/10665/278953Index-TB guidelines: guidelines on extra-pulmonary tuberculosis in India. Country Office for India: World Health Organization; 2016 [accessed 30.09.19].
[19] Department of Health, Republic of South Africa. National tuberculosis management guidelines 2014, <https://health-e.org.za/2014/06/10/guidelines-national-tuberculosis-management-guidelines/>; 2014 [accessed 30.09.19].
[20] Ministry of Health, The Republic of Uganda. A manual for management of tuberculosis and leprosy in Uganda, <http://library.health.go.ug/publications/tuberculosis/manual-management-tuberculosis-and-leprosy-uganda>; 2017 [accessed 30.09.19].
[21] Luo M, Wang W, Zeng Q, Luo Y, Yang H, Yang X. Tuberculous meningitis diagnosis and treatment in adults: a series of 189 suspected cases. Exp Ther Med 2018;16 (3):2770−6.
[22] National Institute for Health and Care Excellence. Tuberculosis, NICE guideline [NG33], <https://www.nice.org.uk/guidance/ng33/chapter/Recommendations#active-tb>; 2019 [accessed 30.09.19].
[23] Jullien S, Ryan H, Modi M, Bhatia R. Six months therapy for tuberculous meningitis. Cochrane Database Syst Rev 2016;9 CD012091.
[24] Thwaites GE, Lan NT, Dung NH, Quy HT, Oanh DT, Thoa NT, et al. Effect of antituberculosis drug resistance on response to treatment and outcome in adults with tuberculous meningitis. J Infect Dis 2005;192(1):79−88.
[25] Heemskerk D, Day J, Chau TT, Dung NH, Yen NT, Bang ND, et al. Intensified treatment with high dose rifampicin and levofloxacin compared to standard treatment for adult patients with tuberculous meningitis (TBM-IT): protocol for a randomized controlled trial. Trials 2011;12:25.
[26] Sun F, Ruan Q, Wang J, Chen S, Jin J, Shao L, et al. Linezolid manifests a rapid and dramatic therapeutic effect for patients with life-threatening tuberculous meningitis. Antimicrob Agents Chemother 2014;58(10):6297−301.
[27] Svensson EM, Svensson RJ, Te Brake LHM, Boeree MJ, Heinrich N, Konsten S, et al. The potential for treatment shortening with higher rifampicin doses: relating drug exposure to treatment response in patients with pulmonary tuberculosis. Clin Infect Dis 2018;67(1):34−41.
[28] Liu Y, Pertinez H, Ortega-Muro F, Alameda-Martin L, Harrison T, Davies G, et al. Optimal doses of rifampicin in the standard drug regimen to shorten tuberculosis treatment duration and reduce relapse by eradicating persistent bacteria. J Antimicrob Chemother 2018;73(3):724−31.
[29] Boeree MJ, Heinrich N, Aarnoutse R, Diacon AH, Dawson R, Rehal S, et al. High-dose rifampicin, moxifloxacin, and SQ109 for treating tuberculosis: a multi-arm, multi-stage randomised controlled trial. Lancet Infect Dis 2017;17 (1):39−49.
[30] Thwaites GE, Duc Bang N, Huy Dung N, Thi Quy H, Thi Tuong Oanh D, Thi Cam Thoa N, et al. The influence of HIV infection on clinical presentation, response to treatment, and outcome in adults with tuberculous meningitis. J Infect Dis 2005;192(12):2134−41.
[31] Choudhri SH, Hawken M, Gathua S, Minyiri GO, Watkins W, Sahai J, et al. Pharmacokinetics of antimycobacterial drugs in patients with tuberculosis, AIDS, and diarrhea. Clin Infect Dis 1997;25(1):104−11.
[32] Gurumurthy P, Ramachandran G, Hemanth Kumar AK, Rajasekaran S, Padmapriyadarsini C, Swaminathan S, et al. Malabsorption of rifampin and isoniazid in HIV-infected patients with and without tuberculosis. Clin Infect Dis 2004;38 (2):280−3.
[33] McIlleron H, Rustomjee R, Vahedi M, Mthiyane T, Denti P, Connolly C, et al. Reduced antituberculosis drug concentrations in HIV-infected patients who are men

or have low weight: implications for international dosing guidelines. Antimicrob Agents Chemother 2012;56(6):3232–8.

[34] Torok ME, Yen NT, Chau TT, Mai NT, Phu NH, Mai PP, et al. Timing of initiation of antiretroviral therapy in human immunodeficiency virus (HIV)—associated tuberculous meningitis. Clin Infect Dis 2011;52(11):1374–83.

[35] Denti P, Martinson N, Cohn S, Mashabela F, Hoffmann J, Msandiwa R, et al. Population pharmacokinetics of rifampin in pregnant women with tuberculosis and HIV coinfection in Soweto, South Africa. Antimicrob Agents Chemother 2015;60 (3):1234–41.

[36] Bandoli G, Palmsten K, Forbess Smith CJ, Chambers CD. A review of systemic corticosteroid use in pregnancy and the risk of select pregnancy and birth outcomes. Rheum Dis Clin North Am 2017;43(3):489–502.

[37] Hiruy H, Rogers Z, Mbowane C, Adamson J, Ngotho L, Karim F, et al. Subtherapeutic concentrations of first-line anti-TB drugs in South African children treated according to current guidelines: the PHATISA study. J Antimicrob Chemother 2015;70(4):1115–23.

[38] Mukherjee A, Velpandian T, Singla M, Kanhiya K, Kabra SK, Lodha R. Pharmacokinetics of isoniazid, rifampicin, pyrazinamide and ethambutol in HIV-infected Indian children. Int J Tuberc Lung Dis 2016;20(5):666–72.

[39] Savic RM, Ruslami R, Hibma JE, Hesseling A, Ramachandran G, Ganiem AR, et al. Pediatric tuberculous meningitis: model-based approach to determining optimal doses of the anti-tuberculosis drugs rifampin and levofloxacin for children. Clin Pharmacol Ther 2015;98(6):622–9.

[40] Thwaites G, Fisher M, Hemingway C, Scott G, Solomon T, Innes J. British Infection Society guidelines for the diagnosis and treatment of tuberculosis of the central nervous system in adults and children. J Infect 2009;59 (3):167–87.

[41] WHO. Treatment of tuberculosis in children: rapid advice. Geneva: World Health Organization; 2010, Available from: https://clinicaltrials.gov/ct2/show/NCT02581527 [accessed 30.09.19].

[42] van Toorn R, Schaaf HS, Laubscher JA, van Elsland SL, Donald PR, Schoeman JF. Short intensified treatment in children with drug-susceptible tuberculous meningitis. Pediatr Infect Dis J 2014;33(3):248–52.

[43] The PIH guide to the medical management of multidrug-resistant tuberculosis. 2nd ed. In: Partners in health. Boston, MA: USAID TB CARE II, <https://www.pih.org/practitioner-resource/pih-guide-to-the-medical-management-of-multidrug-resistant-tuberculosis-2nd>; 2013 [accessed 30.09.19].

[44] WHO. WHO consolidated guidelines on drug-resistant tuberculosis treatment. Geneva: World Health Organization; 2019, Available from: https://www.who.int/tb/publications/2019/consolidated-guidelines-drug-resistant-TB-treatment/en/ [accessed 30.09.19].

[45] Prasad K, Singh MB, Ryan H. Corticosteroids for managing tuberculous meningitis. Cochrane Database Syst Rev 2016;4 CD002244.

[46] Donovan J, Phu NH, Mai NTH, Dung LT, Imran D, Burhan E, et al. Adjunctive dexamethasone for the treatment of HIV-infected adults with tuberculous meningitis (ACT HIV): study protocol for a randomised controlled trial. Wellcome Open Res 2018;3:31.

[47] van Toorn R, Solomons R. Update on the diagnosis and management of tuberculous meningitis in children. Semin Pediatr Neurol 2014;21(1):12–18.

[48] Mai NT, Dobbs N, Phu NH, Colas RA, Thao LT, Thuong NT, et al. A randomised double blind placebo controlled phase 2 trial of adjunctive aspirin for tuberculous meningitis in HIV-uninfected adults. ELife 2018;7.

[49] Schoeman JF, Springer P, van Rensburg AJ, Swanevelder S, Hanekom WA, Haslett PAJ, et al. Adjunctive thalidomide therapy for childhood tuberculous meningitis: results of a randomized study. J Child Neurol 2004;19:250–7.
[50] Soria J, Metcalf T, Mori N, Newby RE, Montano SM, Huaroto L, et al. Mortality in hospitalized patients with tuberculous meningitis. BMC Infect Dis 2019;19(1):9.
[51] Donovan J, Figaji A, Imran D, Phu NH, Rohlwink U, Thwaites GE. The neurocritical care of tuberculous meningitis. Lancet Neurol 2019;18(8):771–83.
[52] Murthy JM. Tuberculous meningitis: the challenges. Neurol India 2010;58 (5):716–22.
[53] Mai NT, Thwaites GE. Recent advances in the diagnosis and management of tuberculous meningitis. Curr Opin Infect Dis 2017;30(1):123–8.
[54] Acosta-Lara P, Varon J. Therapeutic hypothermia in sepsis: to use or not to use? Am J Emerg Med 2013;31(2):381–2.
[55] Misra UK, Kalita J, Bhoi SK, Singh RK. A study of hyponatremia in tuberculous meningitis. J Neurol Sci 2016;367:152–7.
[56] Rivkees SA. Differentiating appropriate antidiuretic hormone secretion, inappropriate antidiuretic hormone secretion and cerebral salt wasting: the common, uncommon, and misnamed. Curr Opin Pediatr 2008;20(4):448–52.
[57] Davis A, Meintjes G, Wilkinson RJ. Treatment of tuberculous meningitis and its complications in adults. Curr Treat Options Neurol 2018;20(3):5.
[58] Misra UK, Kalita J, Kumar M. Safety and efficacy of fludrocortisone in the treatment of cerebral salt wasting in patients with tuberculous meningitis: a randomized clinical trial. JAMA Neurol 2018;75(11):1383–91.
[59] Shu Z, Tian Z, Chen J, Ma J, Abudureyimu A, Qian Q, et al. HIV/AIDS-related hyponatremia: an old but still serious problem. Ren Fail 2018;40(1):68–74.
[60] Rondon-Berrios H, Berl T. Vasopressin receptor antagonists in hyponatremia: uses and misuses. Front Med 2017;4:141.
[61] Misra UK, Kumar M, Kalita J. Seizures in tuberculous meningitis. Epilepsy Res 2018;148:90–5.
[62] Sánchez Fernández I, Gaínza-Lein M, Lamb N, Loddenkemper T. Meta-analysis and cost-effectiveness of second-line antiepileptic drugs for status epilepticus. Neurology 2019;92(20):e2339–48.
[63] Gunawardane N, Fields M. Acute symptomatic seizures and provoked seizures: to treat or not to treat? Curr Treat Option Neurol 2018;20(10):41.
[64] Misra UK, Kalita J. Management of provoked seizure. Ann Indian Acad Neurol 2011;14(1):2–8.
[65] Patwari AK, Aneja S, Chandra D, Singhal PK. Long-term anticonvulsant therapy in tuberculous meningitis—a four-year follow-up. J Trop Pediatr 1996;42(2):98–103.
[66] Sharma D, Shah I, Patel S. Late onset hydrocephalus in children with tuberculous meningitis. J Fam Med Prim Care 2016;5(4):873–4.
[67] Palur R, Rajshekhar V, Chandy MJ, Joseph T, Abraham J. Shunt surgery for hydrocephalus in tuberculous meningitis: a long-term follow-up study. J Neurosurg 1991;74(1):64.
[68] Zhang L, Zhang X, Li H, Chen G, Zhu M. Acute ischemic stroke in young adults with tuberculous meningitis. BMC Infect Dis 2019;19(1):362.
[69] Chatterjee D, Radotra BD, Vasishta RK, Sharma K. Vascular complications of tuberculous meningitis: an autopsy study. Neurol India 2015;63(6):926–32.
[70] Misra UK, Kalita J, Maurya PK. Stroke in tuberculous meningitis. J Neurol Sci 2011;303(1-2):22–30.
[71] Chin JH. Neurotuberculosis: a Clinical Review. Semin Neurol 2019;39:456–61.
[72] Ramachandran R, Muniyandi M, Iyer V, Sripriya T, Priya B, Govindarajan TG. Dilemmas in the diagnosis and treatment of intracranial tuberculomas. J Neurol Sci 2017;381:256–64.

[73] Salway RJ, Sangani S, Parekh S, Bhatt S. Tuberculoma-induced seizures. West J Emerg Med 2015;16(5):625–8.
[74] DeLance AR, Safaee M, Oh MC, Clark AJ, Kaur G, Sun MZ, et al. Tuberculoma of the central nervous system. J Clin Neurosci 2013;20(10):1333–41.
[75] Marais S, Meintjes G, Pepper DJ, Dodd LE, Schutz C, Ismail Z, et al. Frequency, severity, and prediction of tuberculous meningitis immune reconstitution inflammatory syndrome. Clin Infect Dis 2013;56(3):450–60.
[76] Pepper DJ, Marais S, Maartens G, Rebe K, Morroni C, Rangaka MX, et al. Neurologic manifestations of paradoxical tuberculosis-associated immune reconstitution inflammatory syndrome: a case series. Clin Infect Dis 2009;48(11):e96–e107.
[77] Thwaites GE, van Toorn R, Schoeman J. Tuberculous meningitis: more questions, still too few answers. Lancet Neurol 2013;12:999–1010.
[78] Viel-Theriault I, Thibeault R, Boucher FD, Drolet JP. Thalidomide in refractory tuberculomas and pseudoabscesses. Pediatr Infect Dis J 2016;35(11):1262–4.
[79] Ramdasi R, Mahore A, Kawale J, Thorve S. Cerebral venous thrombosis associated with tuberculous meningitis: a rare complication of a common disease. Acta Neurochir 2015;157(10):1679–80.
[80] Bansod A, Garg RK, Rizvi I, Malhotra HS, Kumar N, Jain A, et al. Magnetic resonance venographic findings in patients with tuberculous meningitis: predictors and outcome. Magn Reson Imaging 2018;54:8–14.
[81] Verma R, Lalla R, Patil TB, Tiwari N. A rare presentation of cerebral venous sinus thrombosis associated with tubercular meningitis. BMJ Case Rep 2013. Available from: https://doi.org/10.1136/bcr-2013-009892.
[82] Ramappa V, Aithal GP. Hepatotoxicity related to anti-tuberculosis drugs: mechanisms and management. J Clin Exp Hepatol 2013;3(1):37–49.
[83] Saukkonen JJ, Cohn DL, Jasmer RM, Schenker S, Jereb JA, et al. An official ATS statement: hepatotoxicity of antituberculosis therapy. Am J Respir Crit Care Med 2006;174:935–52.

Further reading

Aulakh R, Chopra S. Pediatric tubercular meningitis: a review. J Pediatr Neurosci 2018;13(4):373–82.

CHAPTER 7

Neurosurgical management of tuberculous meningitis

Peter Ssenyonga
Department of Neurosurgery, Mulago National Referral Hospital, Kampala, Uganda

Key points

- Hydrocephalus is a common complication of tuberculous meningitis.
- CT imaging of the brain is useful for the detection and monitoring of hydrocephalus.
- Medical therapy is recommended for the initial management of hydrocephalus.
- Surgical options for hydrocephalus include ventriculoperitoneal shunting and endoscopic third ventriculostomy.
- Tuberculomas can occur in the brain and spinal cord and usually respond to extended duration treatment with antituberculosis drug regimens.

Introduction

Tuberculosis (TB) continues to be a common communicable disease in low- and middle-income countries [1]. Although global incidence rates of TB are declining slowly, multiple drug-resistant strains of *Mycobacterium tuberculosis* (MTB) and HIV coinfection may be contributing to the perceived rise in central nervous system involvement requiring neurosurgical attention, including tuberculous meningitis (TBM) and tuberculomas. In countries with a high incidence of TB, TBM is typically a disease of young children that develops 3–6 months after primary infection. In countries with a low incidence of TB, TBM more commonly affects adults.

Hydrocephalus

Hydrocephalus is the most frequent complication of TBM occurring in up to 70%–85% of patients [2–7]. Hydrocephalus is the accumulation of cerebrospinal fluid (CSF) in the ventricles resulting from an imbalance between CSF production and absorption, which leads to ventricular

Tuberculous Meningitis
DOI: https://doi.org/10.1016/B978-0-12-818825-5.00007-3

dilatation and raised intracranial pressure (ICP). Hydrocephalus is more common in children and in the later stages of tuberculous meningitis. The most common cause for hydrocephalus is obstruction to CSF flow by tuberculous exudate in the basal cisterns, which leads to a communicating type of hydrocephalus [8–10]. The inflammation of the choroid plexus and ependyma causes an overproduction of CSF in the acute phase of the illness [2] which contributes to the hydrocephalus and raised ICP. The obstructive type of hydrocephalus develops when the fourth ventricular outlets are blocked by exudates and/or leptomeningeal scar tissue or when there is obstruction of the aqueduct either due to a strangulation of the brain stem by exudates or by a subependymal tuberculoma [2].

Radiology

A brain CT scan with and without contrast is our investigation of choice. It is quick, less expensive than MRI, does not require sedation or anesthesia to the patient, and provides sufficient information regarding ventricular size, presence of infarcts, tuberculomas, and exudate in the subependymal layers and basal cisterns. A brain CT scan cannot be used to accurately determine whether the hydrocephalus is communicating or obstructive. This is done by performing an air encephalogram [11,12]. The method we use has been well described by Figaji et al. [11]. A lumbar puncture is done with measuring of opening pressure using a manometer. CSF is taken off and air is injected into the subarachnoid space. The patient is then sat up with the head slightly flexed to allow air to ascend into the cranium. A skull X-ray is then performed to demonstrate the presence of air in the ventricular system if the hydrocephalus is communicating. Certainly, air encephalography is not routinely available in all clinical settings, especially in low-income countries.

Medical management of hydrocephalus

If hydrocephalus is communicating, there is a strong case for medical treatment. The medical management of hydrocephalus secondary to TBM has been validated by a number of studies [10,13]. This requires admission for a prolonged period of time with close monitoring until the ICP normalizes. In our practice, we use the protocol suggested by Schoeman et al. [10,13] that involves (1) anti-tuberculosis chemotherapy, (2) steroids (dexamethasone or prednisone), (3) acetazolamide (100 mg/kg/day),

(4) furosemide (1 mg/kg/day), and (5) serial lumbar punctures with measurement of opening pressures using a simple manometer. Initially, lumbar punctures are performed every day and the duration is increased to every other day and then it increased further until the ICPs normalize. This takes about 2 weeks. The opening pressures should show a downward trend if the medical management is working. A follow-up brain CT is repeated every 3 weeks or sooner if ICPs remain persistently high or there is a deterioration in neurological status. Patients who do not respond the medical treatment are then considered for surgical management [4].

Surgical management of hydrocephalus

For TBM patients who present with a decreased level of consciousness and hydrocephalus, we initially place an external ventricular drain (EVD) for about 48 hours to temporarily divert the CSF. If the Glasgow Coma Scale (GCS) remains low or deteriorates even further with a functioning EVD, the patient is treated conservatively. If the patient's GCS improves, we then perform a permanent method of treating the hydrocephalus. Hydrocephalus of the obstructive type can be treated with either an endoscopic procedure or a shunt. There have been studies to advocate for placing of ventriculoatrial shunts to prevent seeding of the disease to the peritoneal cavity [14,15]. Despite the potential risk of seeding MTB into the peritoneal cavity, from the early 1980s, there has been a shift to ventriculoperitoneal shunting (VPS) as the procedure of choice for patients with hydrocephalus secondary to TBM [2,16,17]. VPS is associated with many complications that have been attributed to the higher protein and cellular content in the CSF leading to more frequent shunt obstruction. When patients with low GCS are shunted, they tend to develop pressure sores along the shunt tract because the patients are not regularly turned, which leads to shunt exposure and infection. However, one needs to be aware that MTB does not colonize the shunt tubing and it is therefore safe to use in the face of active TBM.

There is a place for endoscopic third ventriculostomy (ETV) in patients with obstructive hydrocephalus secondary to TBM. Variable ETV success has been reported by some groups [18–22]. We routinely perform an ETV along with choroid plexus coagulation. Because of the active infection, doing an ETV is a challenge. The tissues are thickened and very friable and the anatomy is sometimes distorted. If the basal cisterns are scarred (which happens in many cases of TBM), ETV will not be successful and VPS will be necessary [23–26]. TBM can sometimes lead to multiloculated hydrocephalus, which requires an endoscopic procedure to communicate the loculations in order to treat the hydrocephalus with a single shunt.

Outcomes

Several factors contribute to a poor outcome in TBM, including cerebrovascular involvement with ensuing brain ischemia, hydrocephalus and raised ICP, direct parenchymal injury, hyponatremia, and seizures [8]. Patients who have a poor GCS at the time of presentation usually have an unfavorable outcome in spite of adequate treatment. Mortality rates range between 10.5% and 57.1% in those with an altered GCS and between 0% and 12.5% for those with normal GCS [27–30].

Tuberculomas

This chapter is not meant to offer an exhaustive details on the neurosurgical management of tuberculomas. They are estimated to constitute 33% of intracranial space-occupying lesions in patients in developing countries [31]. Tuberculomas (single or multiple) may be found in the supratentorial or infratentorial compartment of the brain as well as in the spinal cord. They can exist with or without clinical TBM. CT is reported to have a sensitivity of 100% and specificity of 85.7% and is therefore the investigation of choice. Tuberculomas typically show an isodense or hypodense center with ring enhancement and surrounding edema (see Chapter 3: Clinical presentations and features of tuberculous meningitis) [32]. Sometimes, the lesion may exhibit considerable mass effect mimicking a neoplasm. Tuberculomas demonstrate variable signal intensities on MRI imaging depending on the size and stage of the granuloma [33–35]. Biopsy of the brain is the most accurate method of diagnosis and should be considered in cases without TBM or evidence of TB elsewhere, especially in low incidence countries.

In practice, surgical intervention is rarely required for intracranial tuberculomas. The usual duration of anti-tuberculosis treatment for intracranial tuberculomas and TBM is 12 months with adjunctive corticosteroids for the first 6–8 weeks (see Chapter 6: Treatment guidelines for tuberculosis and tuberculous meningitis). During treatment, enlargement of preexisting tuberculomas or appearance of new tuberculomas have been reported usually between 4 weeks and 17 months [36]. Treatment may need to be continued for up to 34 months to achieve full resolution of the lesions and the decision to stop treatment should be based on the radiological responses of the tuberculomas [37–40]. If the tuberculomas persist or are in the posterior fossa causing hydrocephalus, a case can be made for their surgical resection.

References

[1] WHO. Global tuberculosis report. Geneva, Switzerland: World Health Organization; 2018. <https://www.who.int/tb/publications/global_report/en/> [accessed 11.07.19].
[2] Rajshekhar V. Management of hydrocephalus in patients with tuberculous meningitis. Neurol India 2009;57:368–74.
[3] Cropper MR, Schulder M, Sharan AD, Cho ES. Central nervous system tuberculosis: medical management and surgical indications. Surg Neural 1995;44:378–85.
[4] Donald PR, Schoeman JF. Perspective tuberculous meningitis. N Engl J Med 2004;351:1719–20.
[5] Van Well GT, Paes BF, Terwee CB, Springer P, Roord JJ, Donald PR, et al. Twenty years of pediatric tuberculous meningitis: a retrospective cohort study in the western cape of South Africa. Pediatrics 2009;123:e1–8.
[6] Ozates M, Kemaloglu S, Gurkan F, Ozkan U, Hosoglu S, Simsek MM. CT of the brain in tuberculous meningitis. A review of 289 patients. Acta Radiol 2000;41:13–17.
[7] Thwaites GE, Macmullen-Price J, Tran TH, Pham PM, Nguyen TD, Simmons CP, et al. Serial MRI to determine the effect of dexamethasone on the cerebral pathology of tuberculous meningitis: an observational study. Lancet Neurol 2007;6:230–6.
[8] Figaji AA, Fieggen AG. The neurosurgical and acute care management of tuberculous meningitis: evidence and current practice. Tuberculosis 2010;90:393–400.
[9] Dastur DK, Manghani DK, Udani PM. Pathology and pathogenetic mechanisms in neurotuberculosis. Radiol Clin North Am 1995;33:733–52.
[10] Schoeman J, Donald P, van Zyl L, Keet M, Wait J. Tuberculous hydrocephalus: comparison of different treatments with regard to ICP, ventricular size and clinical outcome. Dev Med Child Neurol 1991;33:396–405.
[11] Figaji AA, Fieggen AG, Peter JC. Air encephalography for hydrocephalus in the era of neuroendoscopy. Childs Nerv Syst 2005;21:559–65.
[12] Bruwer GE, Van der Westhuizen S, Lombard CJ, Schoeman JF. Can CT predict the level of CSF block in tuberculous hydrocephalus? Childs Nerv Syst 2004;20:183–7.
[13] Lamprecht D, Schoeman J, Donald P, et al. Shunting in childhood tuberculous meningitis. Br J Neurosurg 2001;15:119–25.
[14] Bhagwati SN. Ventriculoatrial shunt in tubercular meningitis with hydrocephalus. J 8Neurosurg 1971;35:309–13.
[15] Murray HW, Bandstetter RD, Levyne MH. Ventriculoatrial shunt for hydrocephalus complicating tuberculous meningitis. Am J Med 1981;70:895–8.
[16] Sil K, Chatterjee S. Shunting in tuberculous meningitis: a neurosurgeon's nightmare. Childs Nerv Syst 2008;24:1029–32.
[17] Agrawal D, Gupta A, Mehta VS. Role of shunt surgery in pediatric tubercular meningitis with hydrocephalus. Indian Pediatr 2005;42:245–50.

[18] Figaji AA, Fieggen AG, Peter JC. Endoscopic third ventriculostomy in tuberculous meningitis. Childs Nerv Syst 2003;19:217–25.
[19] Figaji AA, Fieggen AG, Peter JC. Endoscopy for tuberculous hydrocephalus. Childs Nerv Syst 2007;23:79–84.
[20] Singh D, Sachdev V, Singh AK, Sinha S. Endoscopic third ventriculostomy in post-tubercular meningitic hydrocephalus: a preliminary report. Minim Invasive Neurosurg 2005;48:47–52.
[21] Husain M, Jha DK, Rastogi M, Husain N, Gupta RK. Role of neuroendoscopy in the management of patients with tuberculous meningitis hydrocephalus. Neurosurg Rev 2005;28:278–83.
[22] Jonathan A, Rajshekhar V. Endoscopic third ventriculostomy for chronic hydrocephalus following tuberculous meningitis. Surg Neurol 2005;63:32–4.
[23] Marano PJ, Stone SSD, Mugamba J, Ssenyonga P, Warf EB, Warf BC. Reopening of an obstructed third ventriculostomy: long-term success and factors affecting outcome in 215 infants. J Neurosurg Pediatr 2015;15:399–405.
[24] Stone SS, Warf BC. Combined endoscopic third ventriculostomy and choroid plexus cauterization as primary treatment for infant hydrocephalus: a prospective North American series. J Neurosurg Pediatr 2014;14:439–46.
[25] Warf BC, Campbell JW, Riddle E. Initial experience with combined endoscopic third ventriculostomy and choroid plexus cauterization for post-hemorrhagic hydrocephalus of prematurity: the importance of prepontine cistern status and the predictive value of FIESTA MRI imaging. Childs Nerv Syst 2011;27:1063–71.
[26] Warf BC, Kulkarni AV. Intraoperative assessment of cerebral aqueduct patency and cisternal scarring: impact on success of endoscopic third ventriculostomy in 403 African children: clinical article. J Neurosurg Pediatr 2010;5:204–9.
[27] Palur R, Rajshekhar V, Chandy MJ, Joseph T, Abraham J. Shunt surgery for hydrocephalous in tubercular meningitis: a long-term follow-up study. J Neurosurg 1991;74:64–9.
[28] Roy TK, Sircar PK, Chandar V. Ventriculoatrial shunt in the management of tuberculous meningitis. Indian Pediatr 1979;16:1023–7.
[29] Bullock MR, Van Dellen JR. The role of cerebrospinal fluid shunting in tuberculous meningitis. Surg Neurol 1982;18:274–7.
[30] Gelabert M, Castro-Gago M. Hydrocephalus and tubercular meningitis in children. Childs Nerv Syst 1988;4:268–70.
[31] Hejazi N, Hassler W. Multiple intracranial tuberculomas with atypical response to tuberculostatic chemotherapy. Acta Neurochir 1997;139:194–202.
[32] Lwakatare FA, Gabone J. Imaging features of brain tuberculoma in Tanzania: case report and literature review. Afr Health Sci 2003;3:131–5.
[33] Andronikou S, Wieselthaler N. Imaging for tuberculosis in children. In: Schaaf HM, Zumla AI, Grange JM, Yew WW, Pai M, Raviglione MC, et al., editors. Tuberculosis. A comprehensive clinical reference. Amsterdam: Elsevier; 2009. p. 262–96.
[34] Trivedi R, Saksena S, Gupta RK. Magnetic resonance imaging in central nervous system tuberculosis. Indian J Radiol Imaging 2009;19:256–65.
[35] Wasay M, Kheleani BA, Moolani MK, Zaheer J, Pui M, Hasan S, et al. Brain CT and MRI findings in 100 consecutive patients with intracranial tuberculoma. J Neuroimaging 2003;13:240–7.
[36] Shah I, Shetty NS. Duration of anti-tuberculous therapy in children with persistent tuberculomas. SAGE Open Med Case Rep 2019;7 2050313X18823092.
[37] Pauranik A, Behari M, Maheshwari MC. Appearance of tuberculoma during treatment of tuberculous meningitis. Jpn J Med 1987;26:332–4.

[38] Gupta M, Bajaj BK, Khwaja G. Paradoxical response in patients with CNS tuberculosis. J Assoc Physicians India 2003;51:257–60.
[39] Afghani B, Lieberman JM. Paradoxical enlargement or development of intracranial tuberculomas during therapy: case report and review. Clin Infect Dis 1994;19:1092–9.
[40] Poonnoose SI, Rajshekhar V. Rate of resolution of histologically verified intracranial tuberculomas. Neurosurgery 2003;53:873–9.

CHAPTER 8

Evidence gaps and future directions

Jerome H. Chin
Department of Neurology, NYU Langone Health, New York, NY, United States

Is a tuberculosis-free world possible [1–3]? Perhaps not given the widespread prevalence of active and latent tuberculosis and demographic shifts occurring throughout the world. The battle to lower the burden of tuberculosis must prioritize active case-finding in high-prevalence populations and the rapid diagnosis and treatment of symptomatic individuals presenting to health-care facilities in all settings. Development of improved point-of-care diagnostics that can be implemented at low cost with low technical and power requirements will greatly aid these efforts. Evidence-based treatment guidelines as well as more effective drug combination regimens are needed for tuberculous meningitis. Finally, there is a need to improve the clinical training of health workers, especially hospitalists, neurologists, and infectious diseases consultants, in the diagnosis and treatment of meningitis caused by different pathogens, including *Mycobacterium tuberculosis*. Future research of tuberculous meningitis should address the following areas that currently lack reliable data or strong evidence-based recommendations [4]:

1. Global and regional burden of tuberculous meningitis: What are the incidence and mortality rates of tuberculous meningitis in different regions of the world estimated from population-based epidemiological data?
2. Diagnosis of tuberculous meningitis: Does serial sampling and testing of cerebrospinal fluid (three or more samples) increase the detection rate of tuberculous meningitis using Xpert MTB/RIF Ultra or other rapid diagnostic tests? Can next generation sequencing technologies improve treatment outcomes for tuberculous meningitis by increasing the rapid detection rate and/or providing faster and more comprehensive drug susceptibility testing of *M. tuberculosis* in cerebrospinal fluid?
3. Treatment regimens for drug-sensitive tuberculous meningitis: Are the standard four drugs used for pulmonary tuberculosis sufficient to

Tuberculous Meningitis
DOI: https://doi.org/10.1016/B978-0-12-818825-5.00008-5

achieve optimal outcomes for patients with tuberculous meningitis? What should be the minimum duration of treatment for tuberculous meningitis?

4. Adjunctive corticosteroids for tuberculous meningitis: Do corticosteroids offer a mortality and/or morbidity benefit for all subgroups of patients with tuberculous meningitis, including children and HIV-infected individuals?
5. Monitoring during treatment for tuberculous meningitis: How should patients with tuberculous meningitis be monitored and assessed for response and adverse reactions to antituberculosis drugs and for the development of neurological complications?
6. Immune reconstitution inflammatory syndrome: Should initiation of antiretroviral therapy for HIV infection be delayed in patients being treated for tuberculous meningitis?
7. Hydrocephalus: Does earlier and more aggressive treatment of hydrocephalus, for example, external ventricular drainage, reduce mortality and/or severe disability from tuberculous meningitis?

In the future editions of this book, we hope to share the results of research addressing many of these pressing questions. For patients with tuberculous meningitis and their medical providers, any answers will be welcome news.

References

[1] Reid MJA, Arinaminpathy N, Bloom A, Bloom BR, Boehme C, Chaisson R, et al. Building a tuberculosis-free world: the Lancet Commission on tuberculosis. Lancet 2019;393:1331–84.

[2] WHO. The end TB strategy. Geneva, Switzerland: World Health Organization; 2015. <http://www.who.int/tb/strategy/end-tb/en/> [accessed 27.08.19].

[3] WHO. Global tuberculosis report. Geneva, Switzerland: World Health Organization; 2018. <https://www.who.int/tb/publications/global_report/en/> [accessed 26.08.19].

[4] Chin JH. Tuberculous meningitis: a neglected tropical disease? Neurol Clin Pract 2019;9:152–4.

Index

Note: Page numbers followed by "*f*" and "*t*" refer to figures and tables, respectively.

Kolibri 360 GmbH
Ferdinand-Jühlke-Straße 7
99095 Erfurt, Deutschland
produktsicherheit@kolibri360.de